建筑立场系列丛书 No.91

标识与身份

Brand and Identity

贾子光 周荃 | 译

[瑞士] 赫尔佐格和德梅隆建筑事务所等 | 编

大连理工大学出版社

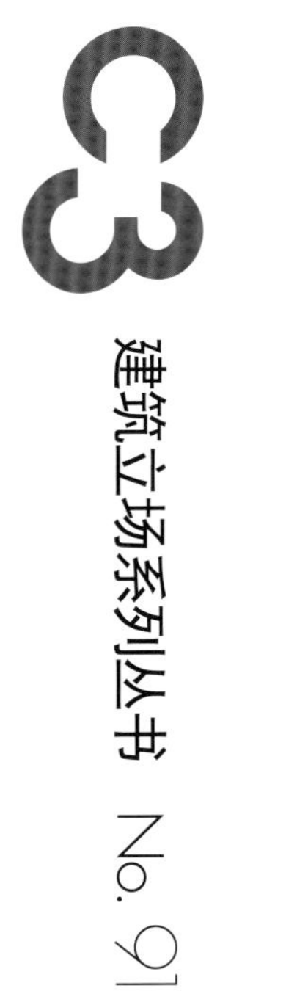

亲近自然——自然栖息地

标识与身份

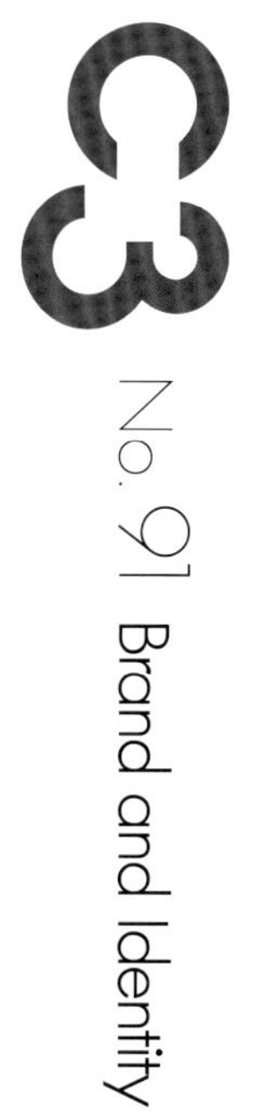

Closing in on Nature - Natural Habitats

Brand and Identity

亲近自然
自然栖息地

Closing in

Natural Habitats

在本文中，作者对位于世界各个角落的四个住宅项目进行了综述，这些项目无一例外地都位于偏远地区。这些项目都是融于大自然之中的，每个都有自己独特的对于场所的描述，而且在设计过程中或之后都遭遇到了不同程度的挑战。总的来说，这些约束条件和设计过程毫无疑问已经在设计中初见成效，因为这些建筑在未来很长一段时间内都将继续融入其引人注目的环境中，并不断地变得更加成熟。当人们生活在自然环境中，作为更加广阔的生态系统的一个"平等"部分时，也许会与大自然产生更深的联系。对往往通过手工方式制作的当地天然材料的利用，增添了当代农村生活的真实性。为了在这种偏远场所创造现代化的生活环境，这些项目完全配备了必需的生活设施。开放的平面设计、毫无杂乱之感的空间成了宁静的观景台和舒适的小憩之处。

Four residential projects, often in remote locations in all corners of the world, are reviewed in this essay. Designed for placement into the very midst of nature, each project has its own unique site-specific narrative, all having encountered a varying degree of challenges during the design process and beyond. Overall, the constraints and processes have altogether unquestionably paid off, as the buildings continue to integrate, mature and morph into their dramatic surroundings long into the future. Perhaps there is a deeper connection with nature, when living in the natural environment as one "equal" part of the wider eco system. Use of locally sourced, natural, often handcrafted materials add to the authenticity of contemporary rural living. The projects come completely equipped with the necessary components for contemporary living in these types of locations. Open plan, clutter free spaces become tranquil viewing ports and comfortable hideaways.

on Nature

亲近自然——自然栖息地

Closing in on Nature - Natural Habitats

Heidi Saarinen

Whilst our cities and urban areas evolve and become denser, faster and more polluted – with work/life balances and other inevitable demands becoming unavoidably extreme, more often, nowadays there seems to be a yearning for "going back to our roots".

As citizens and creators of space; our cities and towns, there is an urgency to insert natural elements, be it materials, green habitable roofs, biophilic courtyards or even levels of vegetation into our buildings and spaces. This is a way of connecting to nature, purifying the air and achieving a different kind of sensory space, where otherwise this does not naturally exist. Looking back into the way our early ancestors created homes in caves; a base made from basic form and material, in natural environments, "to create a sense of comfort and belonging".[1]

At times this approach can appear to be rather frantic, an almost fake belonging to nature. Naturally this is an important part of contemporary design requirements and an aspiration for most, but needs to be sympathetically considered.

Cities are interesting places in their own right; a lifestyle perhaps not negotiable for all. Here though, there are opportunities and cultural chance interactions between people who might not otherwise meet. To chat, to wait, to queue, to shop, to observe – you never know what will happen, whom you may meet. But this existence is not for everyone, and some search for a different kind of place; where the pace is much slower, with a special kind of light, sound and scent. They look for a place, further away, where the seasons may be clearer, or at times harsher; where the topography is a total contrast to that of the city; where the ambience is an invitation for daydream and reflection. Where we may breathe, and just be.

As a child I loved our long family holidays in the wilderness. Modern amenities were deliberately kept to a minimum, for an as-true-to-nature experience as possible. We drove deep into the forest, opened and closed

虽然我们的城市不断地在发展，变得更加密集、节奏更快、污染也更严重——工作与生活之间的平衡，以及其他必然的需求都在不可避免地变得极端，但如今，大家似乎都有一种“返璞归真”的渴望。

作为公民和空间的创造者，我们的城市、城镇迫切需要将自然元素融入我们的建筑和空间之中，无论是材料、适宜居住的绿色屋顶，还是生物庭院，甚至还有植被层。这是一种与自然保持联系的方式，净化空气，获得一种不同的感知空间，否则它就不会自然地存在。回顾我们的祖先早期在洞穴中建造房屋的方式，基础由基本的形式和材料构成，建造在自然环境中，“从而营造舒适感和归属感”。[1] 有时，这种方法会显得相当疯狂，几乎是一种属于大自然的伪装。当然，这是当代设计要求的一个重要部分，也是大多数人的愿望，但需要拿出同理心去进行考虑。城市本身就是很有趣的地方，尽管这里的生活方式可能不是所有人都能接受的。然而在城市里，人们有机会碰面，发生文化上的互动，这些人原本是不会互相认识的。去聊天、去等待、去排队、去购物、去观察——你永远不知道会发生什么，也不知道会遇见谁。但这种生活并不适合所有人，有些人在寻找一种与众不同的地方，那里的生活节奏要慢得多，有一种特殊的光线、声音和气味。他们在寻找一个更遥远的地方，那里的四季可能更加分明，有时气候甚至更严酷，那里的地形与城市完全相反，那里的生活氛围更适于做白日梦或独处反思。这是可以让人自由呼吸的地方，就是这样。

当我还是个孩子的时候，我喜欢全家在荒野中度过悠长的假期。现代化的便利设施被刻意减少至最低的程度，以便于我们都能获得尽可能真实的自然体验。我们把车开进森林深处，打开牛门，又关上，穿过农田和尘土飞扬的乡村道路。洗完桑拿和游泳过后神清气爽，当我们坐在那里吃晚餐时，目送太阳从全景窗的窗口射入波光粼粼的湖面，背后是燃烧的炉火。此时，太多言语都是多余的，我们常常安静地坐着，倾听森林、野生动物发出的声响，和晚风吹拂的声音。这些都给我留下了深刻的印象，我仍然记得其中的细枝末节，并且喜欢这种生活胜过一切。虽然某些神经科学研究表明，其中的一些记忆可能已经发生了变化，尤其是鉴于感知空间的触发可能是由我的快乐记忆所引起的，因此增强了这些特定的记忆或亲近大自然的场景。[2] 有趣的是，自然环境和不同的生活质量如何能如此深刻地影响人类？

the cow gates as we travelled through farmland and dusty rural roads. We watched the sun go down into the reflecting lake from the panoramic windows, as we sat at supper, with the fire in the background, refreshed after the sauna and swim. There was no need for too many words; we often sat in silence, listening to the sounds of the forest, the wildlife and the evening wind. This had a significant impression on me, and I still remember the smallest of details and appreciate it more than anything else. Whilst certain neuro-scientific research may show that some of these memories may have become morphed, as triggers particularly in consideration to sensory space may have been provoked by my happy memories and therefore enhanced these kinds of specific memories or scenarios of closeness to nature.[2] It is interesting how the natural environment and the different quality of life can affect human beings quite so profoundly.

More people are realising the importance of the outdoors, natural wilderness, and silence. Those able to do the shift from urban to natural place will clearly have many benefits to their life, somatic and cognitive wellbeing. But what effect does it have on everyday life and our close understanding of nature? Do we operate differently in a place within nature? If so, what is it that makes people feel, sense and react to their milieus so strongly in the midst of nature? Perhaps those memories are important after all.

One residential building with real sensory and physical connection to the landscape and place is the Bridge House (p.10) by Llama Urban Design located in Ontario, Canada. This building is treating the landscape as one, even though it is topographically challenging with severe level changes within its largely uneven terrain. A bridge above the deepest part of a ravine between two extruded embankments allows access from both sides to the building. The house is constructed from local timber with large areas of glazing, with some smaller windows in interesting and strategic – and otherwise dark – interior places, allowing interaction with the landscape and

越来越多的人开始逐渐意识到户外、自然荒野和寂静的重要性。对于那些能够从城市转移到自然环境中生活的人来说，显然这种改变会对他们的生活、身心健康都有许多好处。但它对日常生活和我们对自然的深入了解有什么影响呢？我们在自然界的某个地方会有不同的做法吗？如果是这样，是什么让人们在大自然中对周围的环境有如此强烈的感受、感知和反应呢？也许终究还是那些记忆很重要吧。

Llama Urban Design事务所在加拿大安大略省设计的桥式住宅（10页），是一座与景观和场所有着真实感知和实际联系的住宅建筑。这座建筑将景观作为一个整体来对待，尽管它在地形上颇具挑战性，在极为不平坦的地势中存在严重的水平高度变化。在两个突出河堤之间的峡谷最深处有一座桥，允许人们从两侧进入建筑。住宅由当地的木材建造而成，开大面积的玻璃窗，在有趣的、有战略意义的（以及其他黑暗的）室内空间内还设计了一些较小的窗户，让人能够与景观互动，并在一天中为空间带来共享的光影。从房子和阳台可以看到令人印象深刻的景观，还有湖泊和远处的荒野；在这里一定有一个地方，适合每一个心情和每一个时刻。室内的摆设比较稀疏，但极有品位，与大自然相互呼应，这一点一直是这个项目室内和室外的设计重点。光线随着季节的变化而变化，成为不断变化的室内自然调色板。

厨房和日常使用区域是开放式的，便于人们相聚和社交，而更多的像卧室这样的私人空间则被布置在空间的两端。储藏室和书籍排列在走廊上。同样，由3r Ernesto Pereira设计的隐形住宅也对现有的景观和周围的地形进行了精心的处理（18页）。这座住宅位于离葡萄牙波尔图35km的内陆地区，业主对感知设计的最细微之处也抱有清晰的想法。为了不干扰环境，也为了能与风景相融而非与之分隔，设计充分尊重了原有的景观，保护环境，尽量做到精简。此外，就像其他设计合理并置身于大自然中的项目一样，这里很显然也是一个独特的场所，给人一种结合了"魔力"和奇迹的感觉。[2]黄昏时分，这栋建筑就像一个灯箱，在附近摇曳的树枝和植被之间发出光芒。白天，建筑结构的形式和框架成为一部从室外取景的机器。未受干扰的树木被精心安排到设计及其附近的景观之中，而远处的森林则成了色彩、运动、阴影和未知环境的背景。

bringing shared light and shadow into the space throughout the day. With impressive views from the house and balcony over the landscape; lake and the wilderness beyond, there is surely a place here for every mood and moment. Interiors are sparse yet tastefully echo the nature that is intentionally central to this inside/outside project. Light changes through the seasons, becoming a constant, naturally shifting colour palette within the interior.
Kitchen and areas of everyday use are open plan, allowing for togetherness and social time, whilst more private spaces such as the bedrooms are placed at opposite ends of the space. Storage and books line the corridors. Similarly, great care was taken of the existing landscape and surrounding terrain in the Cloaked House (p.18) by 3r Ernesto Pereira. Located 35km inland from Porto in Portugal, the client had clear ideas about the smallest of sensory details. The landscape was respected, conserved and "cut around" in order not to disturb the setting too much – and to stay within the scenery rather than outside of it. Again, as with other projects sensibly designed and placed in nature like this, a particular place becomes evident – a sense of "magic" and wonder.[2] At dusk, the building lights up like a lightbox between the swaying branches of the nearby trees and vegetation. Daytime, the form and framing of the structure become a viewfinder from outside in. Undisturbed trees are carefully choreographed into the design and its immediate landscape, whilst the forest beyond act as a backdrop of colour, movement, shade and suspense.
Moenda's House (p.36) by Felipe Rodrigues located in Joanópolis, in the state of São Paulo, Brazil, is positioned on the highest point on a large, sloped site surrounded by protected woodland. The split-level building boasts large open plan spaces, transparency, good ventilation, shade and striking views across the Mantiqueira forest beyond. To the rear of the house, also facing the forest, the owners enjoy an amazing outdoor infinity pool, and a hot tub at one end of the adjacent wooden decking. Concrete exterior patterned screens, allow daylight to enter

由Felipe Rodrigues设计的Moenda住宅（36页）位于巴西圣保罗州的Joanópolis，坐落于一个被保护林地环绕的大斜坡的最高点。这座错层式的建筑拥有巨大的开放空间，视野通透，通风良好，有乘凉地，更能欣赏到远处曼迪奎拉森林那令人惊叹的美景。住宅的后身也面对着森林，业主可以享用一个不可思议的室外无边际泳池，还有在相邻木甲板一端的热水浴缸。拥有混凝土外观花纹的屏风，将入射的阳光变成了各种不同的形状，在一天之中创造了不断变化的光影，也作为永久的通风口使用。起居空间，如休息室、厨房和用餐区位于建筑的主体部分，位于斜坡的较低一层。宽敞的露台提供了通往室外的通道，巧妙地分散了室内外的边界，将景观直接带入室内。建筑师将这一举措描述为该项目的主要目标之一。该项目还有一座独立的客用建筑，包括两间卧室、客厅和设备区。在入口一侧的两个建筑结构之间，设计师还巧妙地布置了一个几乎隐藏的车库。

由Life Style Koubou设计的"一年项目"（28页）以全新的姿态生活在大自然中。这个古怪的项目位于日本猪苗代町的万代山脚下，在建筑设计中引入了浓烈的森林气息。由于该地区在冬季容易受到大雪的影响，因此设计团队不得不创造一个带有结构支柱系统的解决方案，以确保建筑物与地面维持安全的距离。这种做法给建筑带来了一种有趣的堡垒般的感觉，几乎将自然景观完全展现在住户的眼前。

这座住宅被设计成了第二个家，让一家人都能享受乡村的环境，远离他们平常居住的城市。项目有两个木结构体量，分别是主要起居区域和浴室/设备间。一座桥连接了两个空间，正如建筑师所描述的那样，使居住者"在自然环境中享受淳朴和快乐"，用上他们所有的"五种感官"。

"一年"是指施工的时间：从砍伐树木到制造，再到精心地将结构布置到自然场地之中。许多有着80年树龄到90年树龄的古树的

in different shapes, creating changing light and shadow displays throughout the day, also acting as permanent ventilation. The living spaces, such as lounge, kitchen and dining areas are situated in the main block, at the lower level of the slope. Generous terraces give access to the outside, cleverly diffusing the borders between inside and outside bringing the landscape directly into the interior. This move is described by the architects as one of the main objectives of this project. There is a separate guest block with two bedrooms, living and service areas. A cleverly almost hidden garage is placed between the blocks on the entrance side.

One Year Project (p.28) by Life Style Koubou takes a whole new stance to living within nature. Located in Inawashiro, Japan, at the base of Mount Bandai, this quirky project has adopted a strong look of forest into its architecture. As the area is hit by heavy snowfall in winter, the design team had to create a solution of a structural stilt system, to keep the house safely off the ground. This gives the building a playful fortress-like feel, almost surveying the natural landscape, keeping an eye.

The house was designed as a second home, for a family to enjoy the rural setting, away from the city where they normally reside. There are two timber blocks, the main living area and bathrooms/services. A bridge connects the two spaces, making the occupiers "enjoy harshness and happiness in the natural environment" using all their "five senses", as described by the architects.

One Year refers to the time of the process; from cutting the trees to fabrication through to the careful placement of the structure into the natural site. The true form of the many 80-90 year old trees was kept as natural as possible, without special treatments, creating interesting and strong structures. Most importantly, new trees of the same kind replaced the amount of trees that were felled for the building, to grow for future generations and to put back what was taken away. The architects believe that it should be a matter of "borrowing", not just "receiving", or taking from nature.

"Change your brain, your body, your environment in nontrivial ways, and you will change how you experience

真实形态尽可能保持自然的状态，没有特殊的处理，创造了有趣而又坚固的结构。最重要的是，同类型的新树取代了为建房而砍伐的树木，为子孙后代而栽种树木，把被砍伐的树木放回原处。建筑师认为，这应该是"借用"，而不仅仅是"接受"或是从大自然中索取。

"以一种不平凡的方式改变你的大脑、你的身体、你的环境，你就会改变你对这个世界的感受，改变对你来说最有意义的事情，甚至改变你这个人，"马克·约翰逊说。[3]

在交通不便的偏远地区进行设计、规划和施工，面临着诸多挑战。最终的结果可以想象，这是一种森林里的乌托邦式的梦想，天空无限广阔，除了树木和山脉之外，什么都没有，而树木和山脉阻挡了人们想要的阳光。为了实现这个梦想，项目组经历了一段漫长而艰苦的过程。磋商、规范、图纸、计算以及无数的工期，包括"许多参与其中的人所留下的故事"，正如Life Style Koubou在"一年项目"结束之后所描述的那样，在这个由他们创建出来的场所中，每个因素都占据了一个特殊的位置。整个过程的每一部分都与整体相互联系。此外，决定使用当地的材料，并将它们结合到现有的地形当中，相互交织，这在自然项目中是至关重要的，比如我们在本书中所介绍的那些项目。一般情况下，在偏僻的自然地形上施工时，设计概念中最重要的就是，在现存的野生动物和自然栖息地中安静地创造一个野外的场所，不对原有的生态造成干扰。当然，这几乎是不可能的，但必须继续把它作为先决条件。维护建筑、花园和场地是另一项额外的工作，这将成为日常工作的重要组成部分。

生活在接近自然的地方，也许是为了追求更健康、更安宁的生活节奏。这对大多数人来说可能是一种奢侈，不是所有人都能享受到的，然而，可能是一种理想、一个梦想。可以很容易在城市空间、小露台和窗台上创建这类小型自然保护区。它们虽然远远不及我们在这里探讨的那些引人注目、设计细致入微的项目，但很明显，即使是最小的自然景观对我们所有人也都是非常重要的。

your world, what things are meaningful to you, and even who you are", says Mark Johnson.[3]

Designing, planning and constructing in remote, awkward and not easily accessible locations has many challenges. The end result; imaginably a kind of utopian dream in the forest, with the sky as big as it gets, nothing but trees and mountains to get in the way of desired daylight, has undergone a long and strenuous process. Negotiations, regulations, drawings, calculations and the countless stages, including the "many stories by the people involved" in the process, as described by Life Style Koubou after the One Year Project, all have a significantly special part in the place that has been created. Every part of the process is interconnected to the whole. Additionally, decisions to use local materials and to shape and intertwine them into the existing topography are crucial in nature-projects such as those we have covered here. Commonly, the idea when building on natural, remote terrain, importantly; amongst existing wildlife and nature habitats is to quietly create positionality into the wild, without disturbance. Of course this is almost impossible, but must continue to be a prerequisite. Maintaining the building, gardens and site beyond is another "extra" that will be very much part of the daily routine.

Living close to nature is to strive for, for a healthier, more peaceful pace of life perhaps. This may be a luxury for most, not accessible to all, but possibly an ideal; a dream, nevertheless. Mini nature sanctuaries can be created easily in urban spaces, on small terraces and windowsills. Although far from the attractive designed-to-detail projects that have been reviewed here, it is clear that even the smallest bit of nature is important for all of us.

1. A. Betsky, *Landscrapers: Building with the land.* London: Thames and Hudson, 2002
2. S.Robinson & J. Pallasmaa, eds., *Mind in Architecture: Neuroscience, Embodiment, and the Future of Design.* Cambridge: The MIT Press, 2015, p.200
3. Johnson, M. *The Meaning of the Body.* Chicago: University of Chicago Press, 2007, p.82

桥式住宅

Bridge House

Llama Urban Design

这座桥式住宅位于多伦多市以北两小时车程的地方，坐落于安大略省悉尼港的玛丽湖岸边。建筑体量位于峡谷之上，在两棵大枫树之间，对场地的影响是最小的。它坐落在斜坡最陡峭的位置，画了一条38m长的水平线，与景观互为呼应。这种简单的姿态是为了赞美这个场所给人带来的体验，不仅在你穿过房子、越过峡谷的时候，而且当你走到房子下方的滨水区时，就像一个人穿过增强了景观宏伟规模的门廊一样。

这座住宅有两个主立面。其中一个面对着湖，就像面向风景开放的阳台，位于峡谷上方6m处，将使用者置于树冠的高度。而另一个面朝森林，有一个遵循结构形式的巨大洞口，反映了峡谷的自然轮廓。

这座住宅悬挂在一个倒置的V形胶合结构上，它也支撑着外部楼梯，连接内部的社交区域和屋顶平台。建筑整体产生了一种轻盈的感觉，从内部形成了观赏景观的框架，允许外部的视线横穿过住宅。

精心制作的木质饰面只使用当地的木材制造，产生了一种与周围景观共鸣的对比。外部覆盖着水平铺设的未染色雪松壁板，强调了场地体量的方向性。室内覆盖着大面积的枫木胶合板，作为大自然微妙运动的背景或投影屏幕。在这里，树木的运动创造了光影的游戏，与海滩上波浪拍打的声音相得益彰。

The Bridge House is located two hours north of the city of Toronto, on the shores of Mary Lake in Port Sydney, Ontario. Placed over a ravine, between two large maple trees, the volume has a minimal impact on the site. It sits across the steepest part of the slope, drawing a 38-meter-long horizontal line that acts as a counterpoint to the landscape. This simple gesture celebrates the experience of the place, not only as one walks through the house, over and across the ravine, but also when heading down to the waterfront, underneath the house, as one would cross a portico that heightens the monumental scale of the landscape.

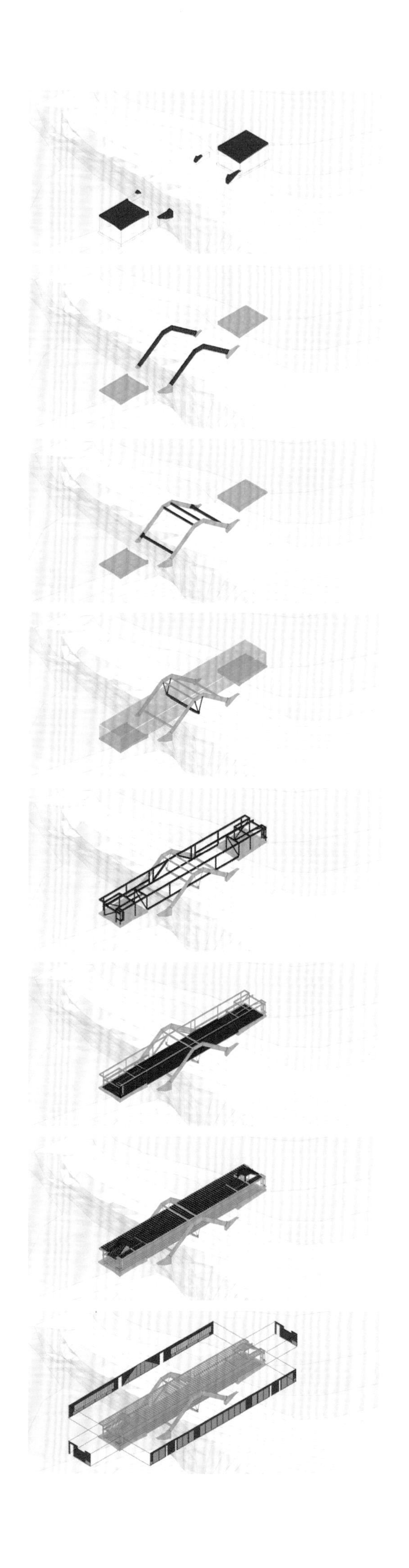

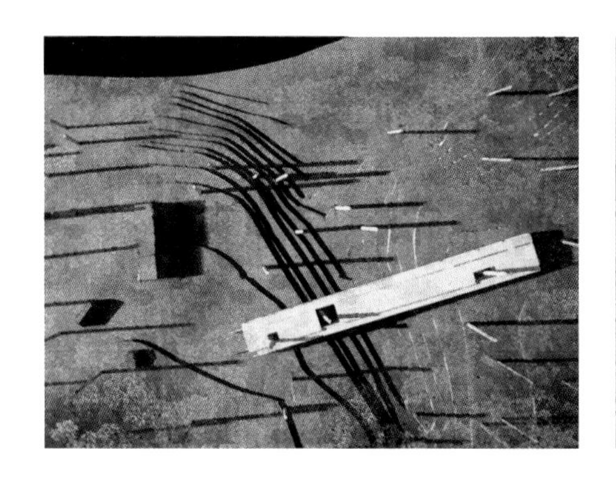

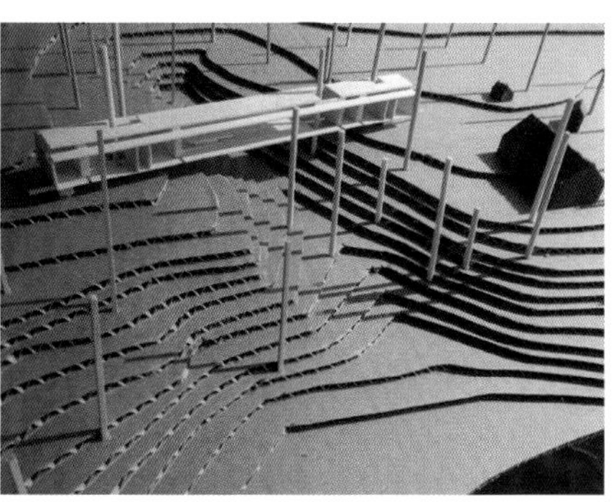

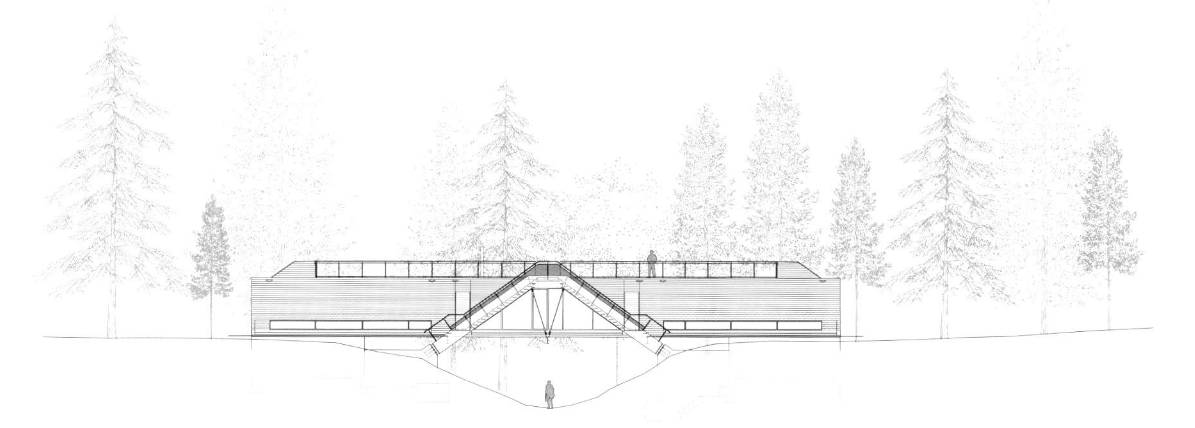

东立面 east elevation

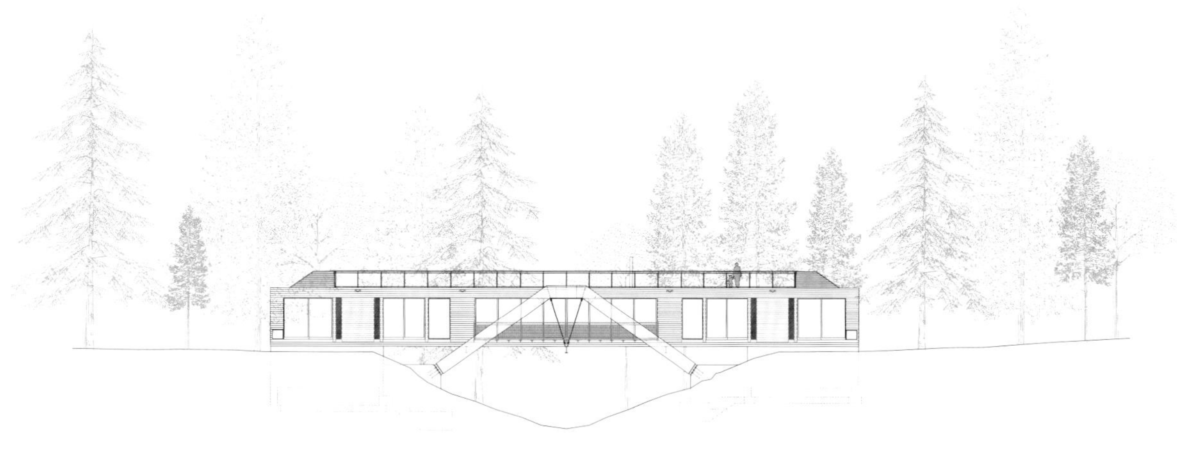

西立面 west elevation

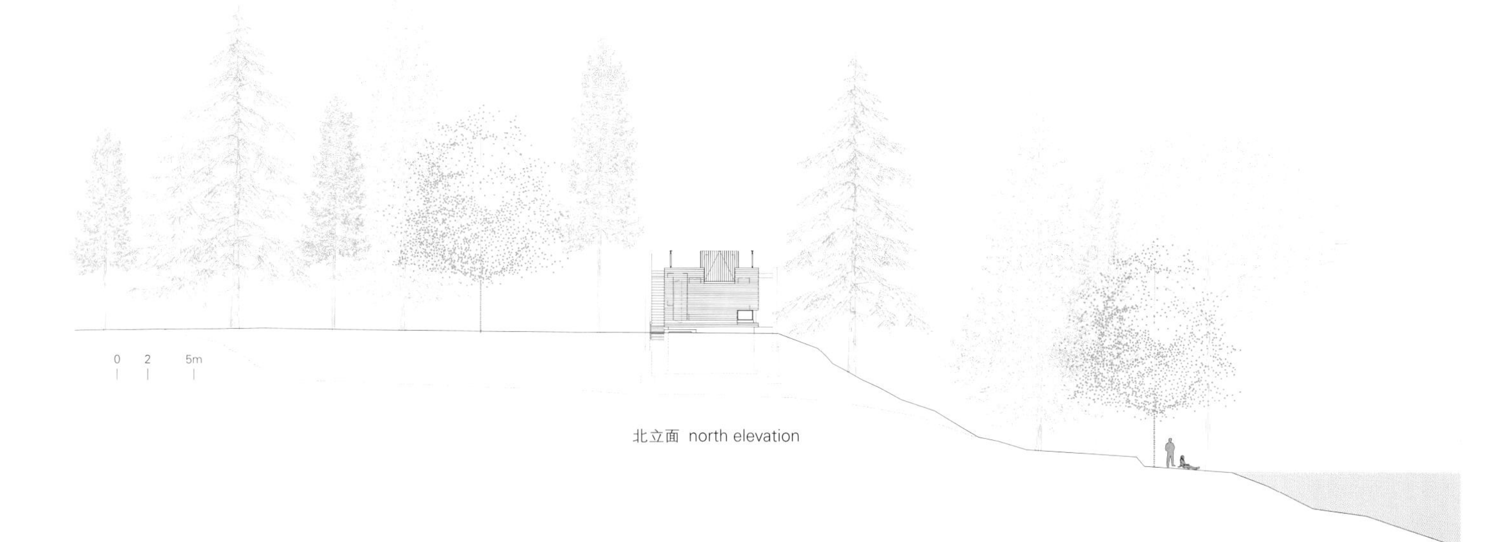

北立面 north elevation

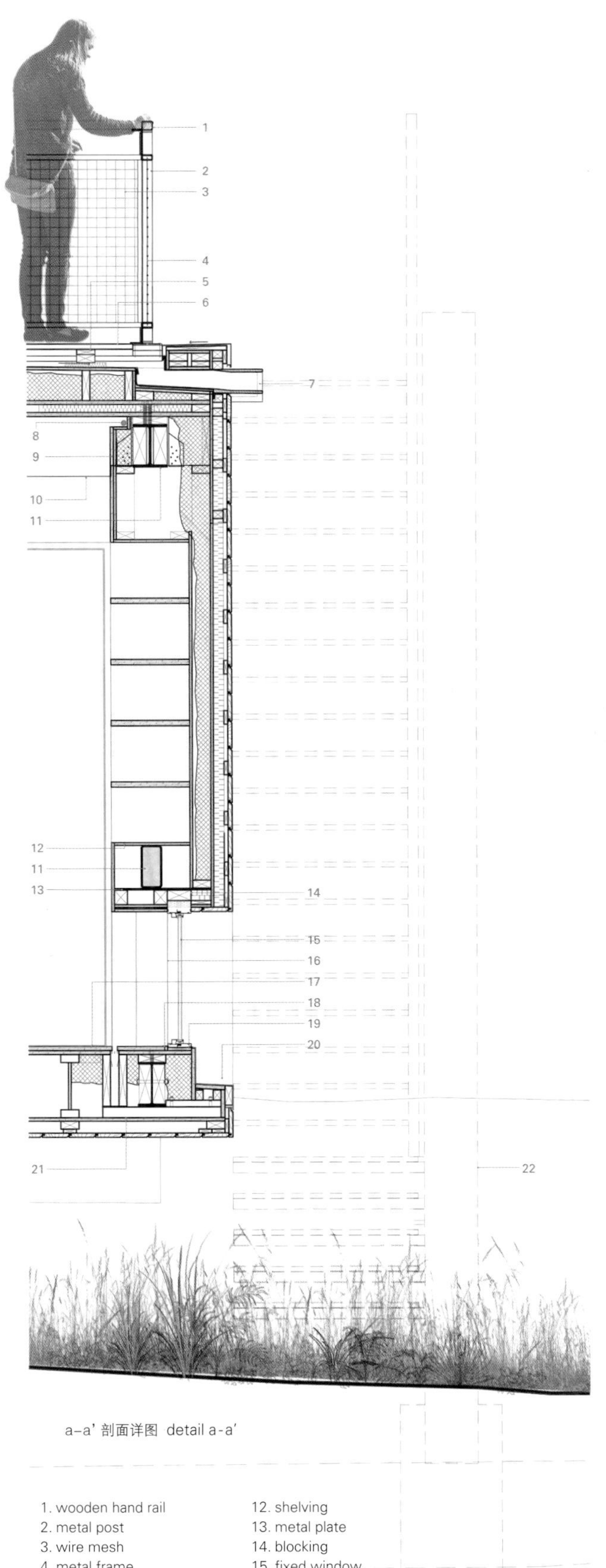

a–a' 剖面详图 detail a-a'

1. wooden hand rail
2. metal post
3. wire mesh
4. metal frame
5. wood block - rigid insulation
6. deck
7. scupper
8. LED lighting
9. maple plywood
10. beam
11. metal beam
12. shelving
13. metal plate
14. blocking
15. fixed window
16. column behind
17. electrical outlet
18. joint with cap
19. support for window sill
20. flashing
21. wood nailer
22. projection showing main beam

bottom of ravine

The house has two main facades. One faces the lake, acting as a sort of balcony that opens up towards the landscape, 6 meters above the ravine, placing the user at the height of the tree canopy. The other faces the forest and has a large opening which follows the form of the structure, mirroring the natural outline of the ravine.

The house hangs from an inverted V-shaped glulam structure that also supports the exterior stairs, connecting the interior social area with the roof deck. The ensemble produces a sense of lightness that frames the landscape from the interior, allowing the exterior to cross through the house.

A carefully crafted wooden finish, using only local woods, produces a contrast that resonates with the surrounding landscape. The exterior is clad in unstained cedar siding that runs horizontally, emphasizing the directionality of the volume on the site. The interior is clad in large panels of maple plywood, acting as a backdrop or projection screen to the subtle movements of nature. Here, the play of light and shadow created by the movement of the trees is complemented by the lapping sound of small waves on the beach.

项目名称：Bridge House
地点：Port Sydney, Ontario, Canada
建筑师：Mariana Leguía, Angus Laurie - Llama Urban Design
项目团队：Mariana Leguía, Angus Laurie, Patrick Webb, Alvaro Rivadeneira
工程师：Mike Feindel - Blackwell Structural Engineers; Gravenhurts Heating & Pluming
用地面积：10,000m² / 建筑面积：230m²
结构：wood and steel hybrid structure
材料：maple ply - interior; unfinished cedar siding - exterior
施工时间：2014—2016
摄影师：©Ben Rahn/A-Frame studio (courtesy of the architect)

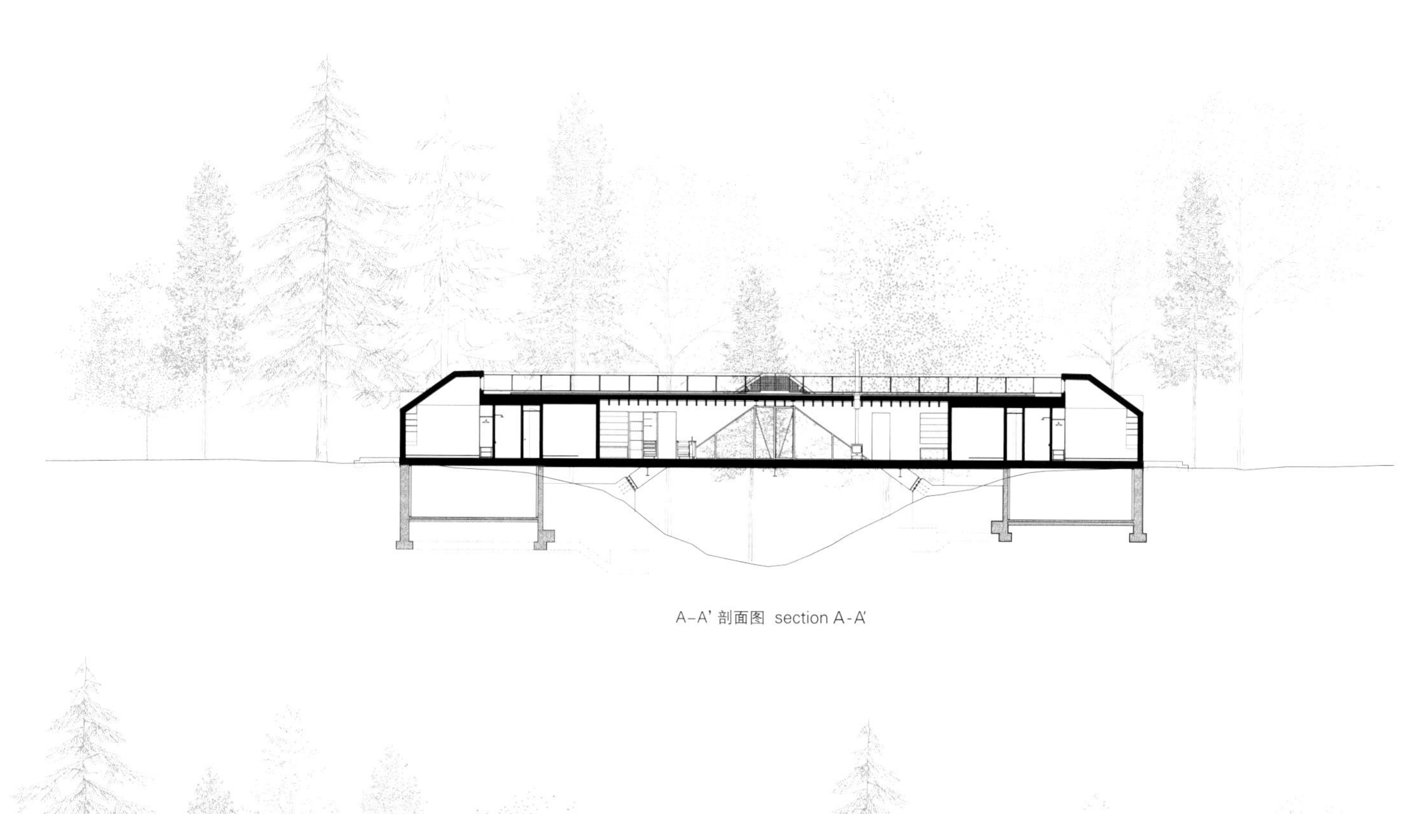

A–A' 剖面图 section A-A'

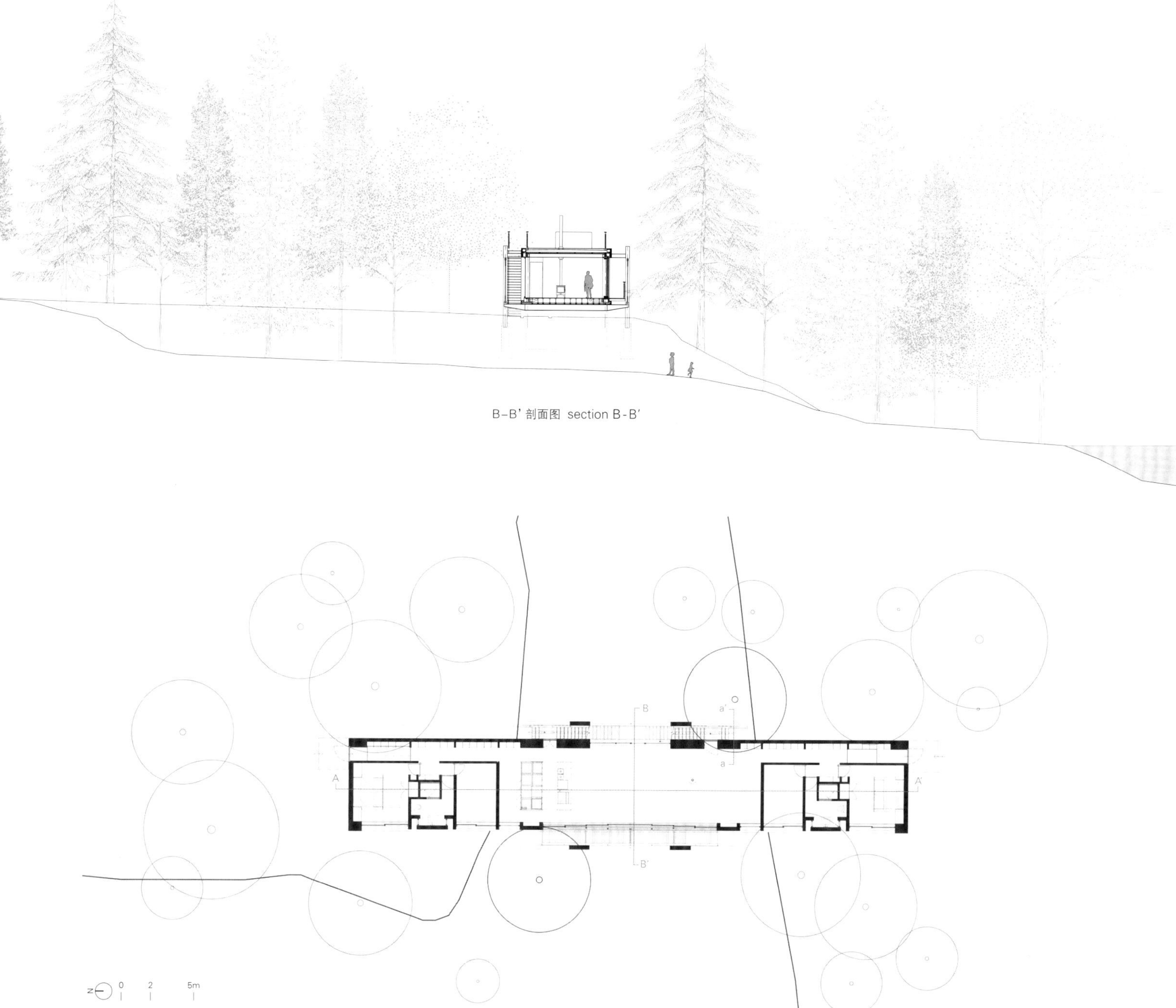

B–B' 剖面图 section B-B'

隐形住宅

Cloaked House

3r Ernesto Pereira

有一片栗树环绕的土地，被一条顺着山坡流下的小溪环绕着，这里坐落着一座隐形住宅，这座建筑的设计显然是受到了场地的启发。它一方面以令人难以置信的姿态融入山中，另外也有宏伟开放的一面，可以让人尽情欣赏大自然的美景。的确，这座房子给我们带来了震撼的体验，在我们思考、行走和生活的过程中留下了印记。最终，它变成了一所住宅，“充满了大胆的感官喜悦，在这所住宅中，你会看到不可思议的事情，处处是奇迹！”正如住宅的设计委托人亚历山德拉·费雷拉所说的那样。

“隐形住宅”的概念来自于对周围自然的田园场景的融合，并进行伪装和改造，而不是简单地强加于其上。这座住宅强调对周围自然环境的保护，为了维护现有的树木而做出适当的“改造”；屋顶和楼板这两个“叶片”一旦遇到树枝就会被刺穿。因而，天井将动态的氛围引入到房子中，最显著的特色是支撑屋顶花园的木柱子排列出了清晰的节奏。住宅的其余部分是透明的玻璃，在这个场景中，玻璃的采用不失为一个成功的解决方案。

夏天，浓密的树叶“吞没”了整个房子，使它变得几乎难以觉察，同时，也保护了室内免受强烈的阳光照射。冬天，落叶树会落叶，树木变得更加显眼，同时，也会邀请阳光进入室内，将房间变得暖洋洋。

很难界定我们是身处室内还是室外，是生活在森林中还是与森林共处，是暴露于大自然之中还是与大自然交流。但毫无疑问，这种模棱两可的设计激发了人们生活在其中的意愿。

On a piece of land surrounded by chestnut trees and encircled by a stream flowing down the hillside, there sits the Cloaked House, clearly inspired by its site. While incredibly merging into the mountain, it has magnificently open side that embraces the view of the natural landscape. Indeed, the house brings sensational experience that leaves their mark as we think, walk and live within it. Fruitfully, it turned into a house "full of daring delight for the senses, where the unlikely happens and magic abounds!", as expressed by the client of the house, Alexandra Ferreira.

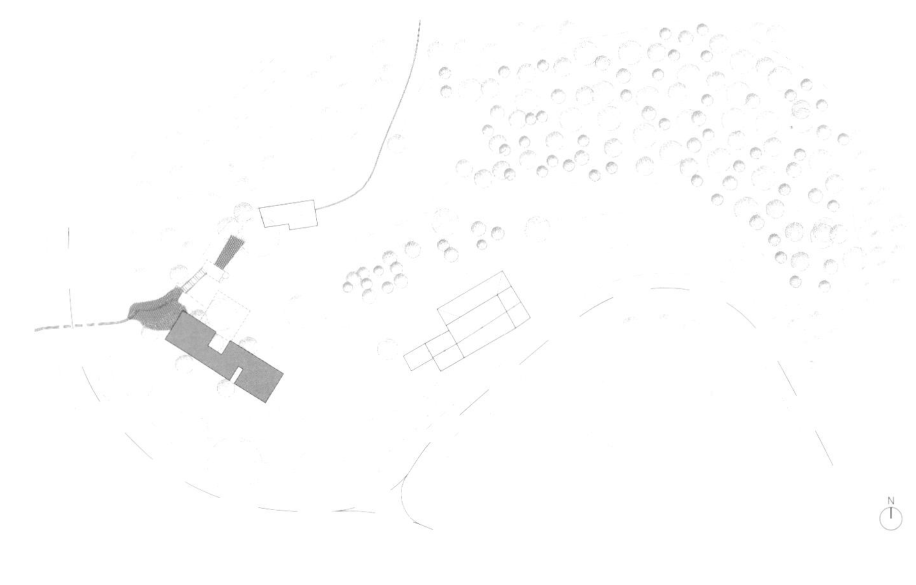

项目名称：Cloaked House
地点：Marco de Canaveses, Portugal
建筑师：3r Ernesto Pereira
总建筑师：Ernesto Pereira
项目建筑师：Ernesto Pereira, Daniela Leitão, Tiago Martins -3r Ernesto Pereira
施工与协调：3r Ernesto Pereira. Arquitetura + (Re)construção
项目团队：Helder Moreira, Joaquim Linhares, José Rocha, Maria José Pereira, Joaquim Pinto, Paulo Silva, Rui Silva, Bruno Moreira, Marco Silva
总面积：140m²
室内面积：90m²
造价：100,000€
施工期间：4 months
竣工时间：2017
摄影师：©João Morgado (courtesy of the architect)

The concept of Cloaked House was derived from the bucolic scenario of blending, camouflaging, transforming the surrounding nature, not to simply impose on top of it. The house emphasizes the conservation of the neighboring nature, thus "transforms" to maintain the existing trees; the two "blades" - the roof and the floor slab – are punctured whenever encountering a tree branch. This has resulted in patios that introduce dynamic atmosphere into the house, which is marked by a well-defined rhythm of the wooden pillars that support the garden-topped roof. The rest of the house is transparent with glass, which was a successful solution in this scenario.

In summer, dense leafiness of the trees "engulfs" the whole house, making it almost imperceptible, and at the same time, protecting the interior from the intense sunlight. In winter, deciduous trees shed their leaves, making it slightly more visible, and simultaneously, inviting sunlight into the house to warm it up.

It is difficult to define whether we are indoors or outdoors, living in the forest or with the forest, exposed to nature or in commune with nature. But definitely, this ambiguity fuels the will to live in it.

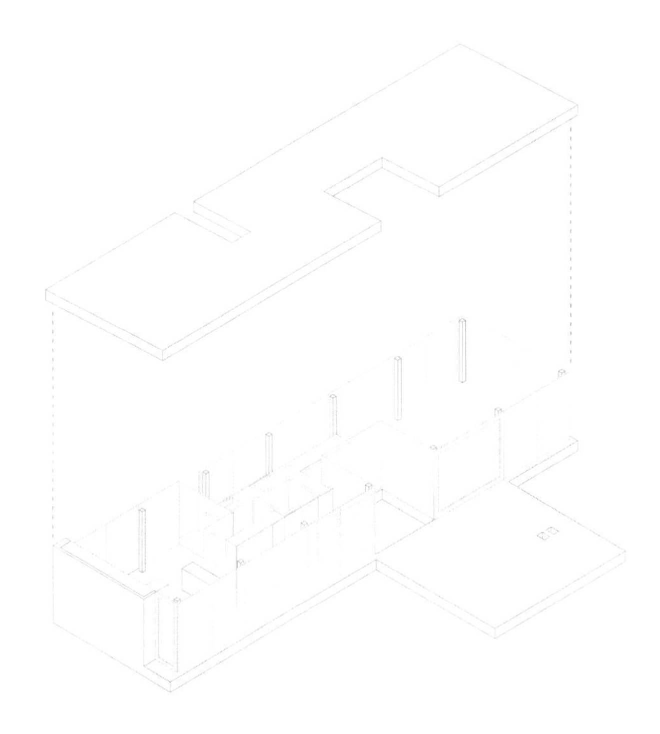

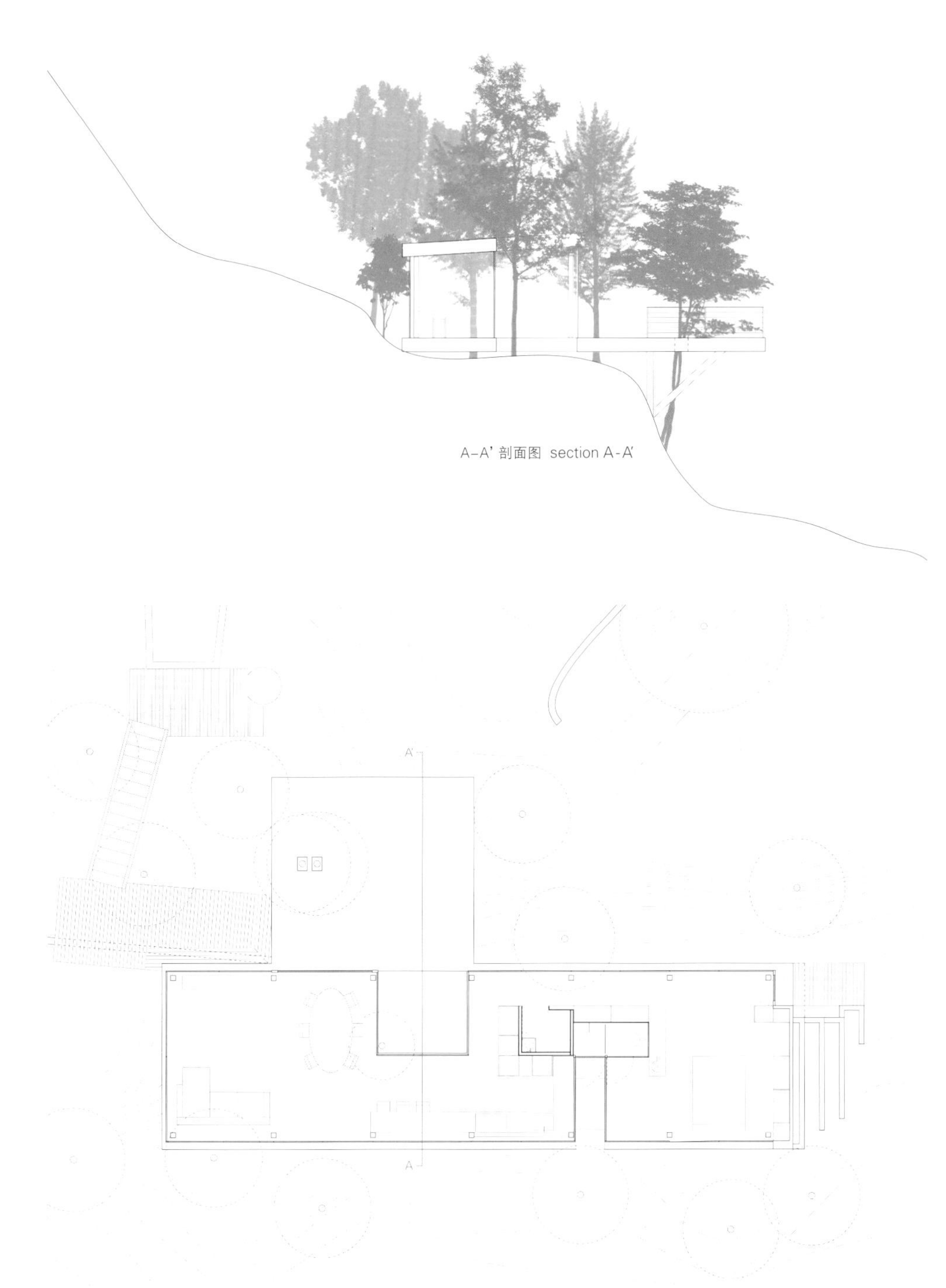

A–A' 剖面图 section A-A'

一年项目

One Year Project

Life Style Koubou

面向自然、适应自然的建筑

该住宅位于日本的万代山脚下，是为一个主要居住在城市的家庭设计的第二个家。由于该地区已被指定为强降雪区域，所以设计包括了一个支柱结构系统，保证了2m的雪荷载阻抗距离。住宅由两个体块组成：一个生活区和一个配备了水暖设备的区域（包括浴室和厨房），各司其职。这两个体块由一座开放的连桥连接，这座连桥能够让居民们尽情地享受大自然之壮美。

顾名思义，"一年项目"的核心思想是与时俱进。从砍树、精细加工木材到材料的组装，整个项目的组装过程在荒野中进行了整整一年。这座住宅的设计核心是保持与大自然的对话，设计和施工过程中的每一步都是对话。砍树一般在冬季进行，以保证获取品质最佳的材料。然后木材经过一段时间的处理，可以保存半年到三年。目前，一个住宅项目平均需要大约4个月的时间，相比之下，一年的项目非常长了；然而，建筑结构的建造——从切割到锯切、加工、运输、装配和风景园林设计——产生了许多精彩的故事，这要感谢参与各个工序的人们。

共有120棵树被用来建造这座建筑，每棵树的树龄都在80年到90年之间，直径约为40cm。木材没有经过特殊处理，以保持其原有的木质特征。木材被切割成非标准尺寸，每一件都是一件精湛的工艺作品，为建筑注入了生命。

虽然"一年工程"是一种按照现代社会规则运作的人造结构，但是，只要有可能，施工工程都会屈服于自然环境。虽然，一般情况下我们不可能将一棵树的每一部分都用来制作建筑材料，但项目顺应自然的灵活性使其得以实现。

树木是一种自然资源，当它们被人类消耗使用之后，就必须重新栽种。这个项目不是获取，而是从环境中借来的。当项目完成后，将种植与项目所使用的树木数量完全相同的树木，以使资源回归自然，为子孙后代积累更多的资源。生长缓慢的树木，它的生命不为时间所俘获；它与人同在，人也与它同在。

Architecture that faces and adjusts to nature

Located at the base of Bandai Mountain, the house is designed as a second home for a family whose primary residence is in the city. As the region has been designated a heavy snowfall area, the design incorporates a stilt structure system to guarantee a snow load resistance of 2 meters. The house is composed of two blocks: a living area, and a plumbing-equipped area (including bathroom and kitchen), each in its own section. The blocks are connected by an open bridge that allows the residents to enjoy the beauty of untamed nature with all their senses.

As the name implies, the "One Year Project" is centered on the idea of moving with the times. The process of putting together the project, from cutting the trees, refining the lumber, and assembling the materials took an entire year in the wilderness. The construction of the house was, at its core, a dialogue with nature, each and every step of the process a conversation. Tree-cutting is generally done in the middle of winter to guarantee the highest-quality material. The lumber is then treated for a period of time that can last from half a year up to three. The average time required for a residential project today is approximately four months, which makes the One Year Project very long in comparison; however, the construction of the structure – from the cutting to the sawing, the processing, the delivery, the assem-

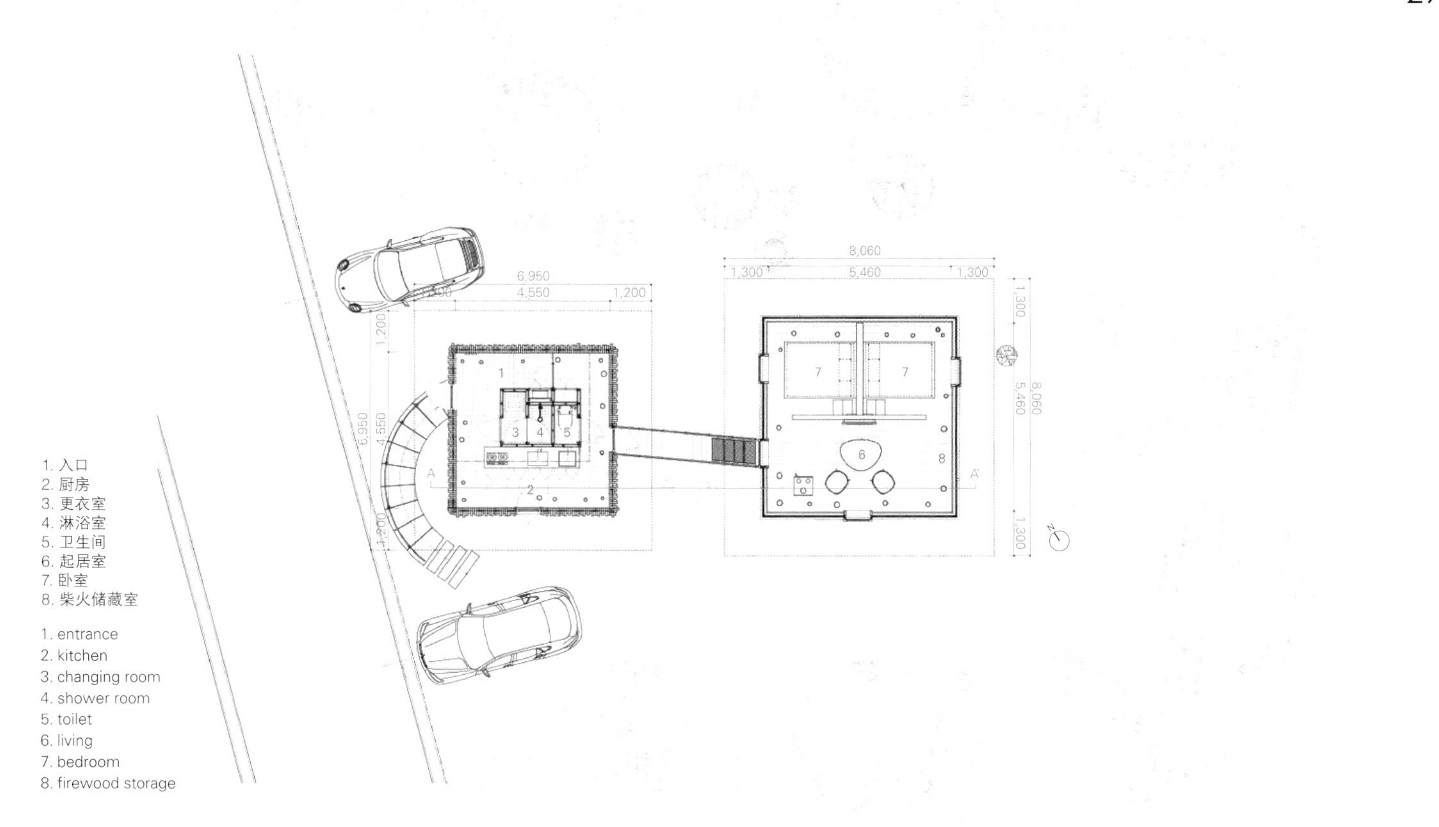
1. 入口
2. 厨房
3. 更衣室
4. 淋浴室
5. 卫生间
6. 起居室
7. 卧室
8. 柴火储藏室
1. entrance
2. kitchen
3. changing room
4. shower room
5. toilet
6. living
7. bedroom
8. firewood storage
6,950
4,550
1,200
8,060
1,300
5,460
1,300

bling, and the landscaping – gave birth many wonderful stories, courtesy of the people involved in each step.
One hundred and twenty trees, each approximately 80 to 90 years old and approximately 40cm in diameter, were used in the construction of this building. The lumber was not specially treated, so as to maintain their original wooden features. The lumber was cut into non-standardized sizes, each piece a work of fine craftsmanship that comes together to breathe life into the buildings.
Although the One Year Project is a manmade structure that functions in accordance with modern society's regulations, the construction project yielded to nature wherever possible. Although it is normally not possible to use every part of a tree when processing it for use in building materials, the flexibility offered by the project's yielding to nature allowed for it to happen.
Trees are a natural resource, which must be replaced and nurtured whenever they are consumed for human use. Rather than taking, this project borrowed from the environment. After the completion of the project, the very same kind of trees used for the project will be planted in order to return the resource to nature and build up more resources for future generations. The life of a slow-growing tree is not held captive by the times; it moves with the people, and the people move with it.

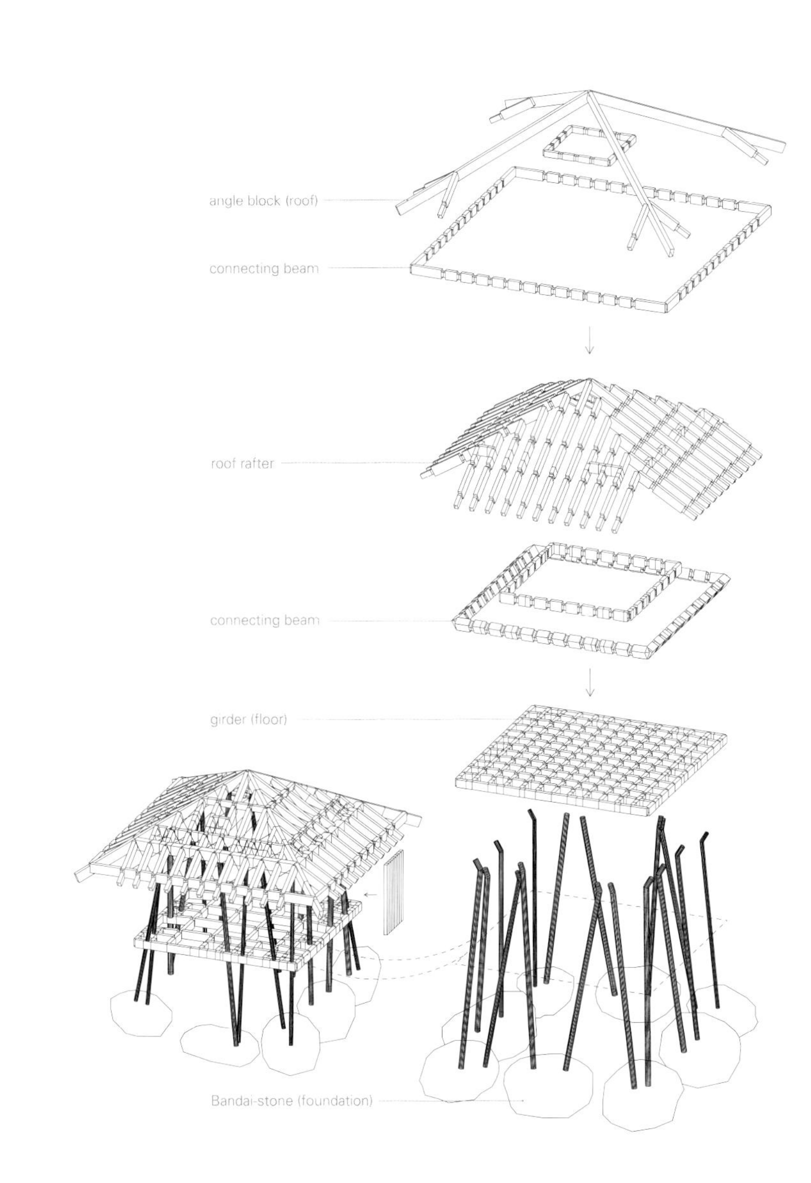

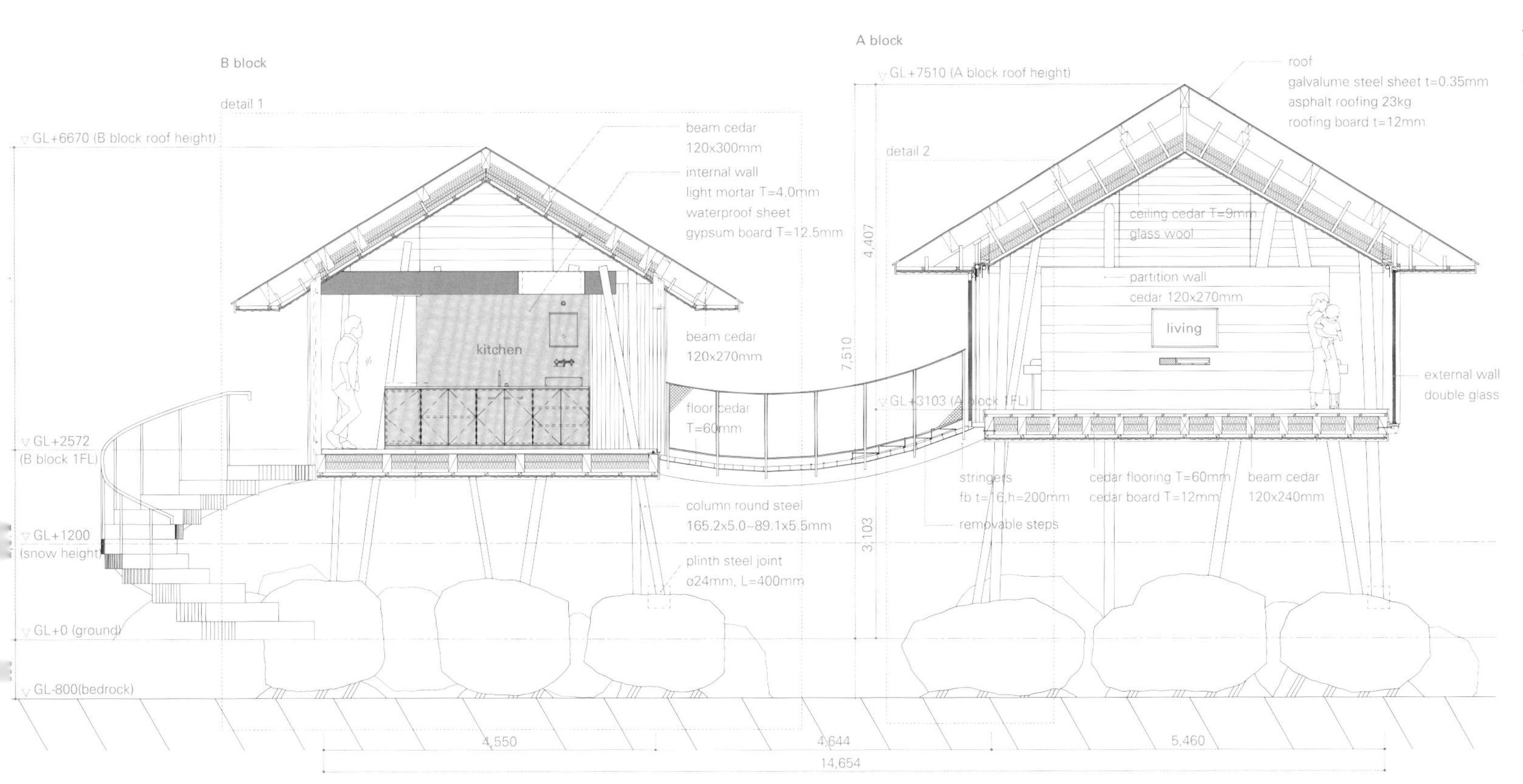

A–A' 剖面图 section A-A'

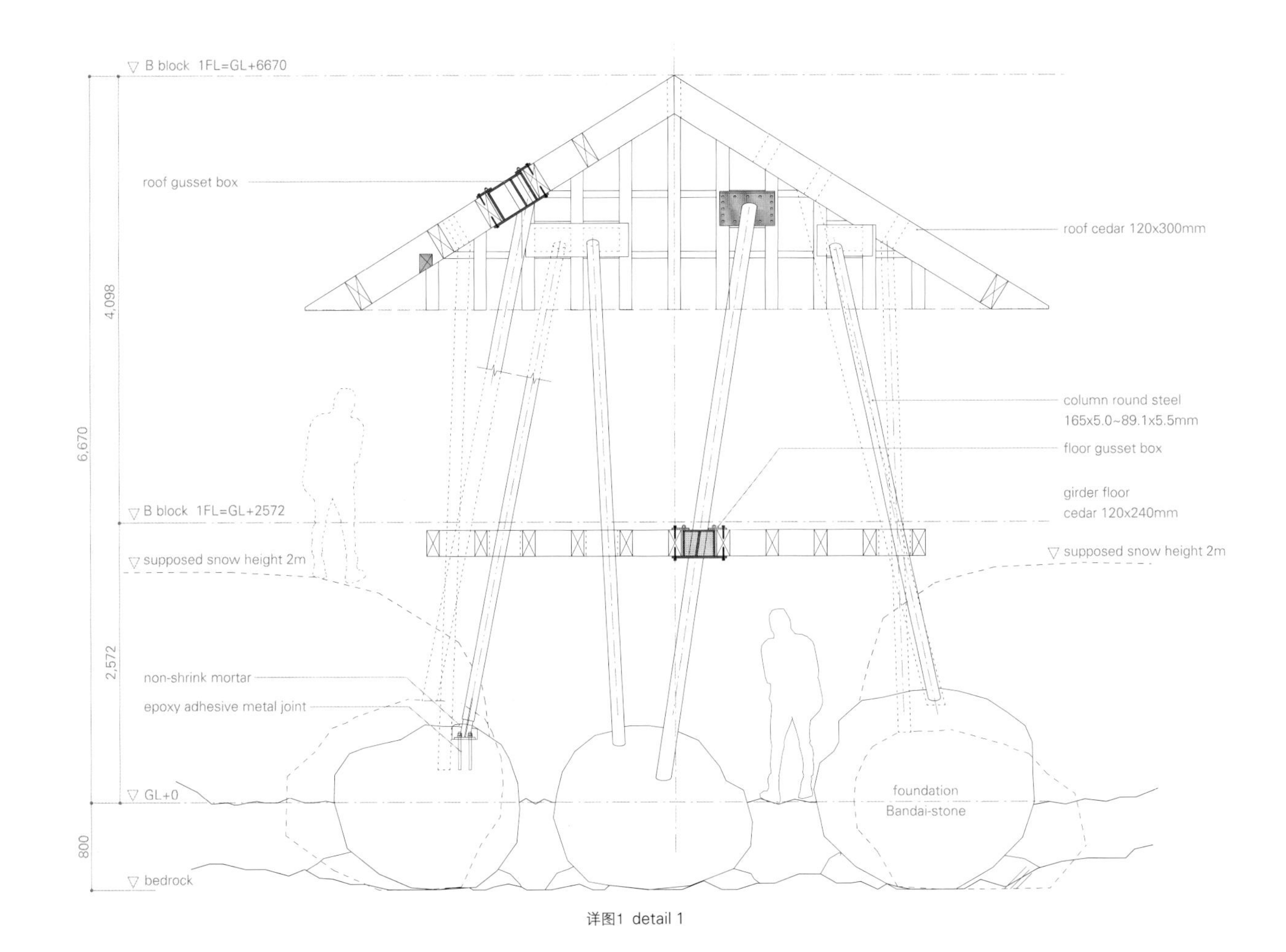

详图1 detail 1

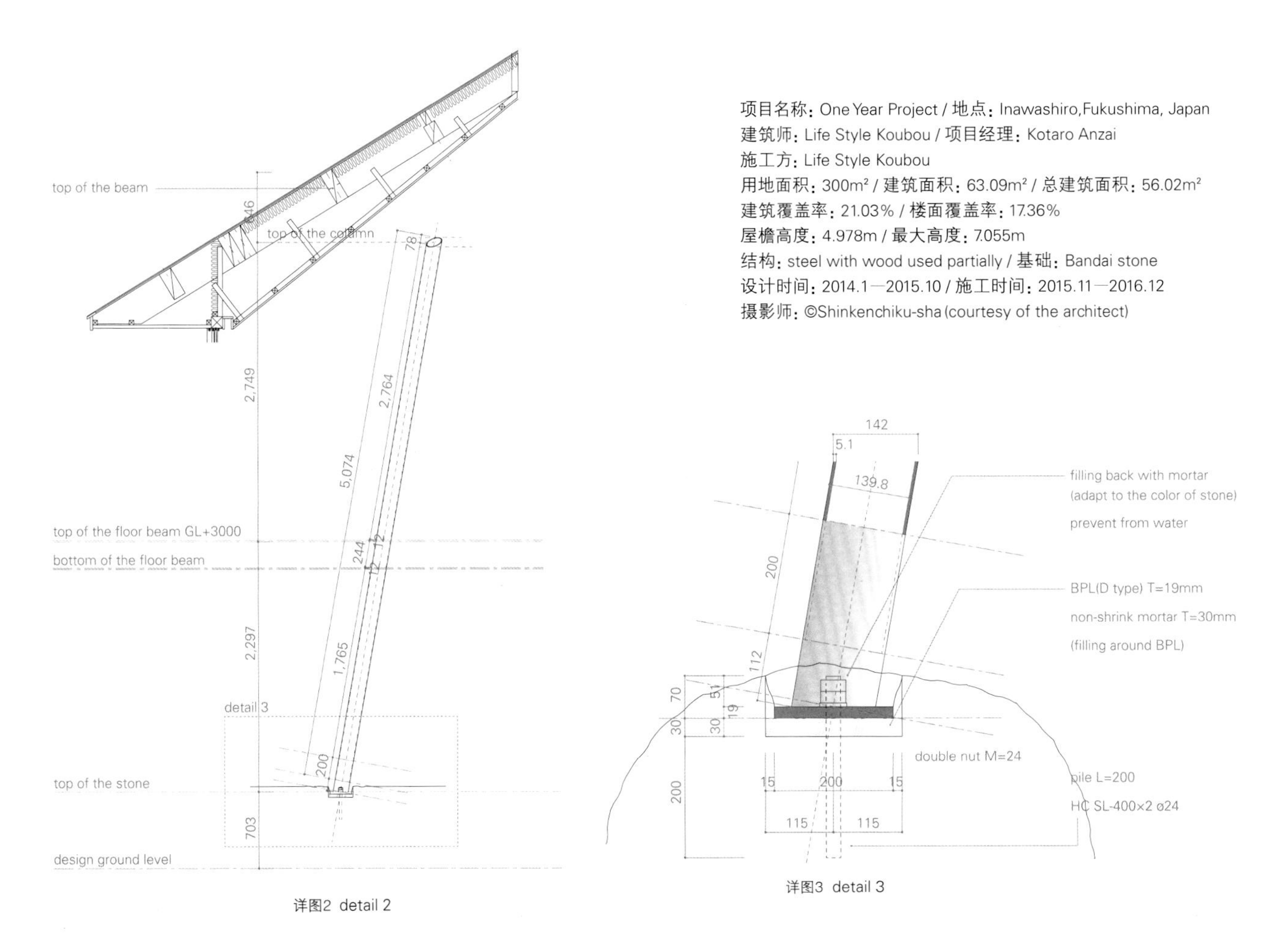

详图2 detail 2

详图3 detail 3

项目名称：One Year Project / 地点：Inawashiro,Fukushima, Japan
建筑师：Life Style Koubou / 项目经理：Kotaro Anzai
施工方：Life Style Koubou
用地面积：300m² / 建筑面积：63.09m² / 总建筑面积：56.02m²
建筑覆盖率：21.03% / 楼面覆盖率：17.36%
屋檐高度：4.978m / 最大高度：7.055m
结构：steel with wood used partially / 基础：Bandai stone
设计时间：2014.1—2015.10 / 施工时间：2015.11—2016.12
摄影师：©Shinkenchiku-sha (courtesy of the architect)

Moenda 住宅

Moenda House

Felipe Rodrigues

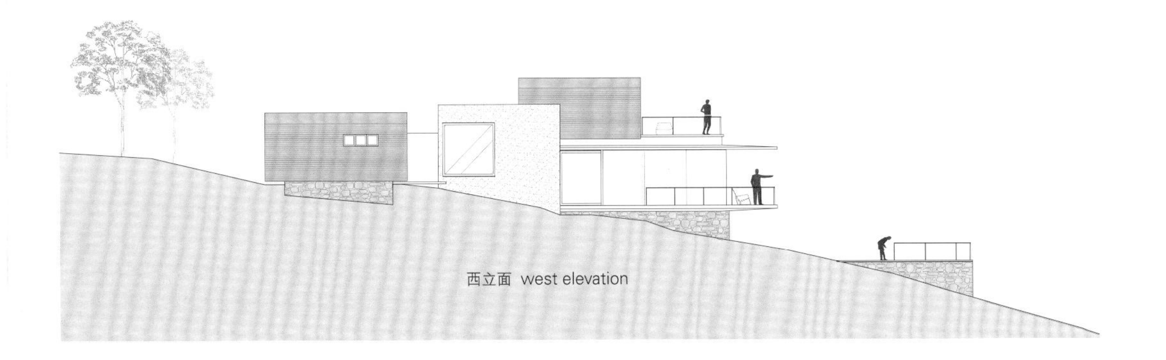

西立面 west elevation

这处位于圣保罗州的占地2万平方米的房产，是最近新建的一个细分农场中的第一个。该地块约有70%为受到保护的原始森林用地，处于开发项目的最高点，前有全景视野，后有Mantiqueira森林保护。

住宅两个独立体量的设计是建筑师对客户要求有独立客房的回应。第一个体量是为实际的业主准备的，将会得到更多的使用，而第二个是为客人准备的。

在主楼较低的一层，在两块钢筋混凝土板之间，是客厅、电视机房、厨房以及露台，露台上覆盖着宽大的悬臂结构，采用玻璃板和落地窗与室内分隔开来。这个决定创造了宽敞的开放空间，保证了充足的通风，更重要的是，达成了项目的一个重要愿望，那就是在建筑内部也能描绘出外面的景观。

在同一体块的中间楼层有两间套房，而上层是一间带有露台和木制平台的小工作室。在这个上层，一堵无窗墙隔开了水泵装置和水箱。无边际泳池面向森林，对房子的主要体量形成了补充。客房则包括两间套房卧室、一个用餐区和设备区。车库位于延长的屋顶板下方，构成了建筑中两个体块之间的入口和交通流线。卧室交通流线中的混凝土屏风的选用尤其值得关注。有了它，才确保了住宅良好的自然采光和永久通风，同时统一了细长的立面，并通过光影营造出妙趣横生的视觉效果。

The 20,000m² property in the state of São Paulo was the first of a recently subdivided farm to be built on. With around 70% occupied by protected virgin forest land, the plot is at the highest point of the development, with a panoramic view of the horizon in front and protected by the Mantiqueira forest in the rear.

The two separate volumes of the house were the architect's response to the client's wish for an independent guest-house. The first, which will be used more often, is for the actual owners, while the second is for guests.

In the main block, at the lower level, between two reinforced concrete slabs, are the sitting-room, TV room and kitchen, as well as the terraces, which are covered by generous cantilevered structures and separated from the inside by glass panes and floor-to-ceiling sliding doors. This

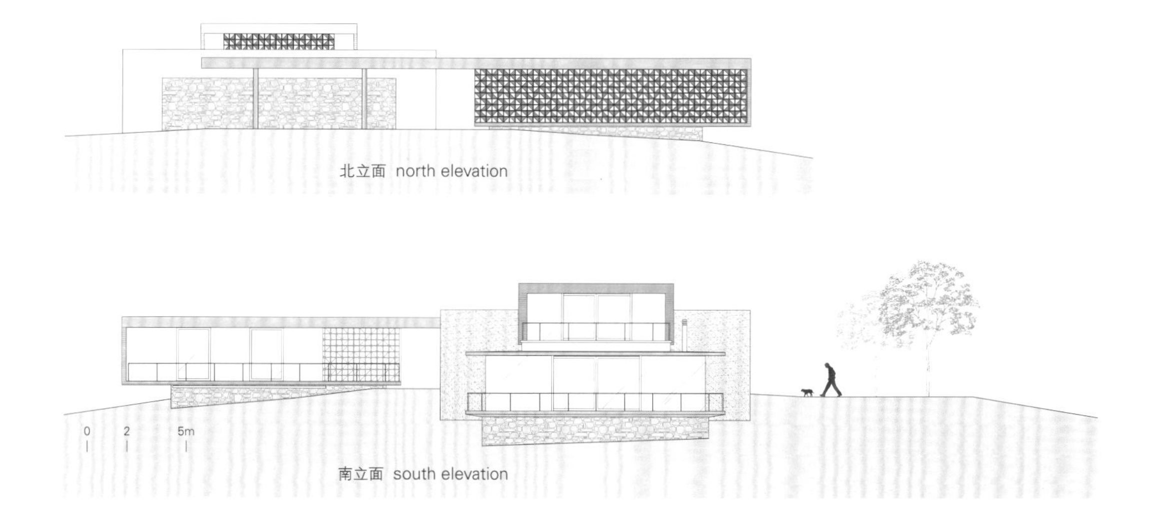

北立面 north elevation

南立面 south elevation

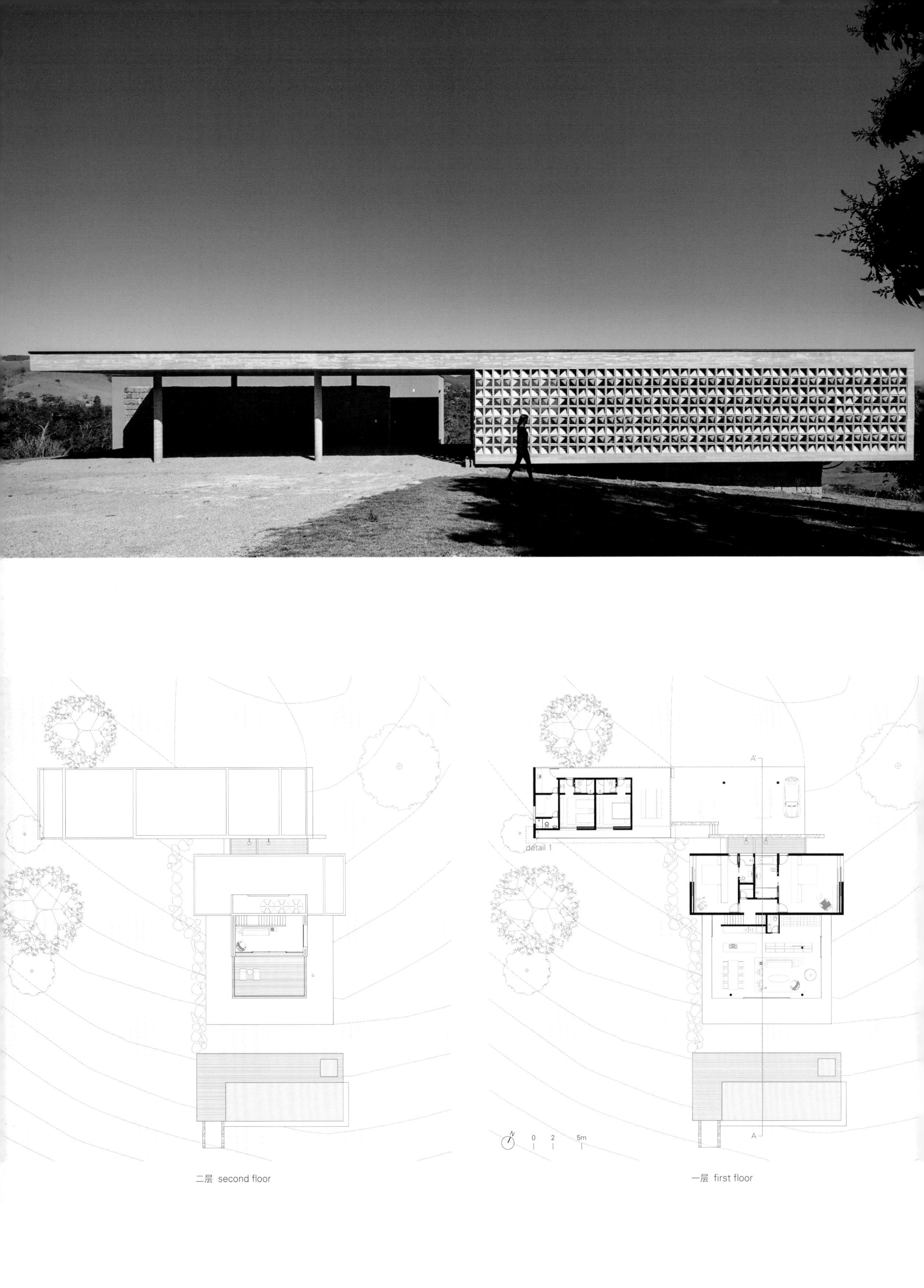

二层 second floor

一层 first floor

0
5
10m

decision has created wide open spaces with plentiful ventilation, and more importantly, enabled one of the project's main aspirations, which was to draw the landscape inside the building.

In the same block, at the mid level, are two en-suite rooms, and at the upper level, a small studio with a terrace and wooden deck. Still on the upper level, a blind wall separates the pump installation and water tanks. The infinity pool faces the forest and complements the main volume of the house.

The guest block houses two en-suite bedrooms, a dining area and the service areas. The garage is located beneath the prolonged roof slab and constitutes the point of access to and circulation between the two blocks in the construction. The choice of concrete screen blocks in the circulation of the bedrooms is worthy of note. This permits natural lighting and permanent ventilation, while unifying the elongated facade and producing a very interesting visual effect with light and shadows.

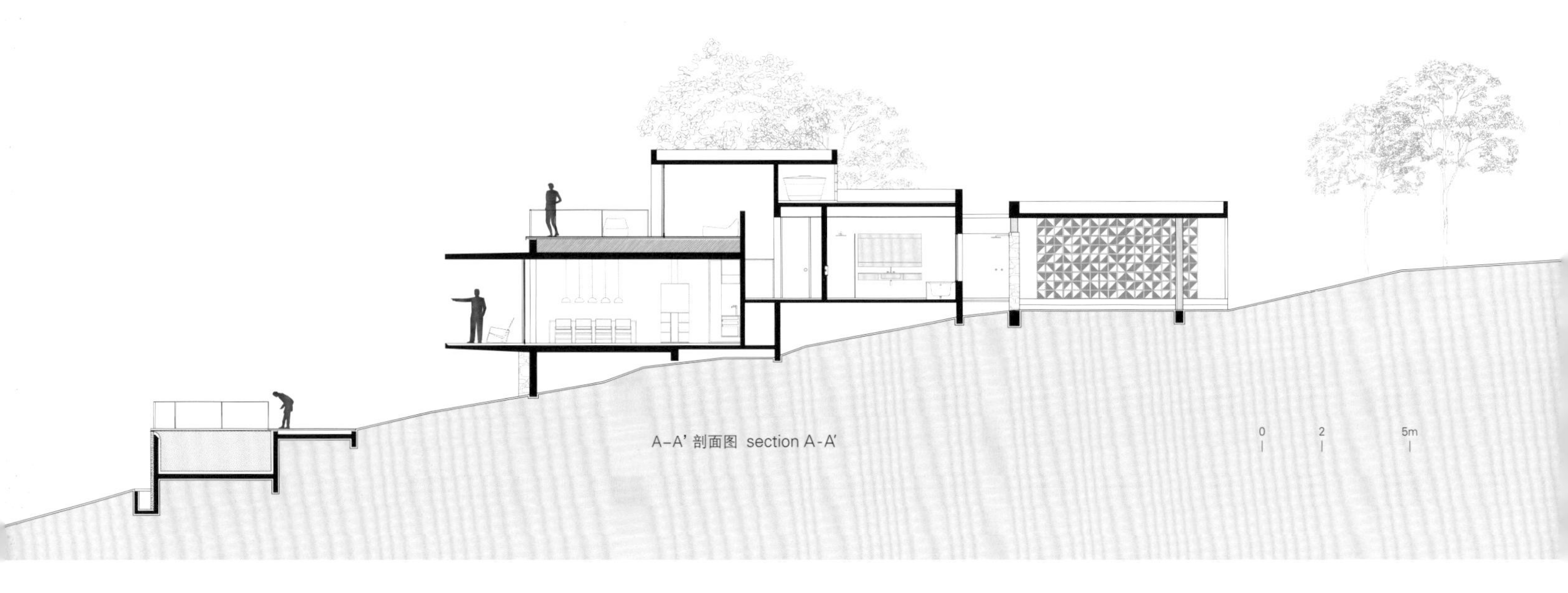

A-A' 剖面图 section A-A'

项目名称：Moenda House / 地点：Moenda-Joanópolis, São Paulo, Brazil / 建筑师：Felipe Rodrigues
混凝土结构：Benedictis / 混凝土顾问：GR Consultoria / 建造商：Marcio Freire / 景观设计师：Bonsai Paisagismo
家具：Baraúna, Decameron, Phenícia e Felipe Rodrigues / 建筑面积：510m² / 竣工时间：2017
摄影师：©Pedro Vannucchi (courtesy of the architect)

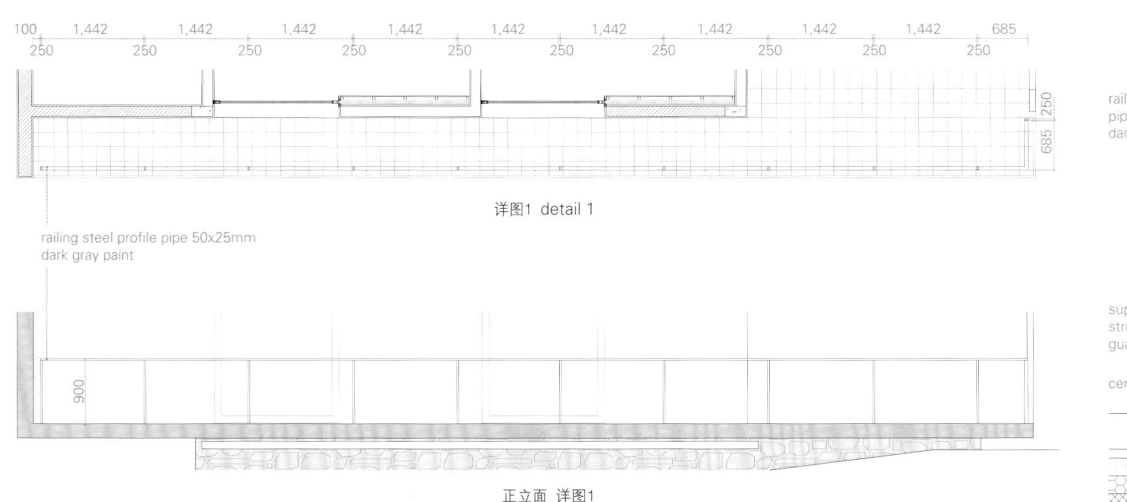

详图1 detail 1

正立面_详图1
front elevation_detail 1

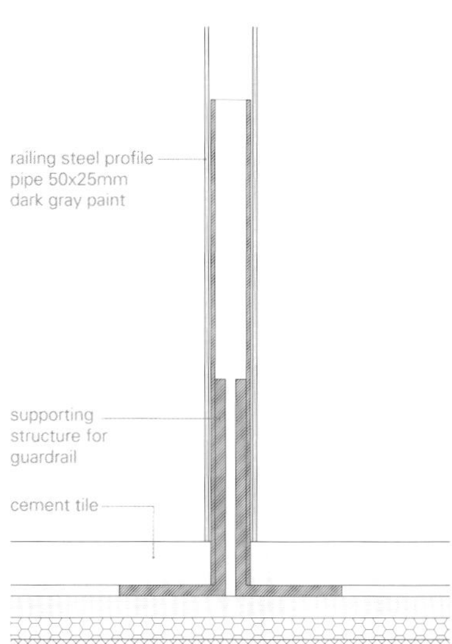

护栏_详图1
guardrail_detail 1

Brand and

标识与身份

本书所介绍的项目说明了如何利用建筑设计来传递关于公司及其品牌的强烈信息。这里展示的几栋建筑可以看作是企业形象的最终体现，企业形象代表了每家公司的特征。从化妆品到轮胎，从纺织到银行公司，这些跨国公司和他们委托的建筑师一起工作来创造建筑，这些建筑物不仅在其所处的城市环境中确认了它们的存在，而且也体现了这些公司所秉持的使命、精神和价值观。在这一部分中，我们将看到不同的建筑师如何利用建筑设计来实现每个公司的身份。

The projects featured in this book illustrate how architecture can be used to convey a strong message about a company and its brand. The several buildings presented here can be seen as the ultimate manifestation of the corporate image that characterises each company. From cosmetics to tires and from textile to banking corporations, these global companies and the architects they employed worked together to create buildings that not only affirm their physical presence in the urban contexts in which they sit, but they also render the mission, ethos and values for which these companies stand. In this section we will see how different architects used architecture to materialise each company's identity.

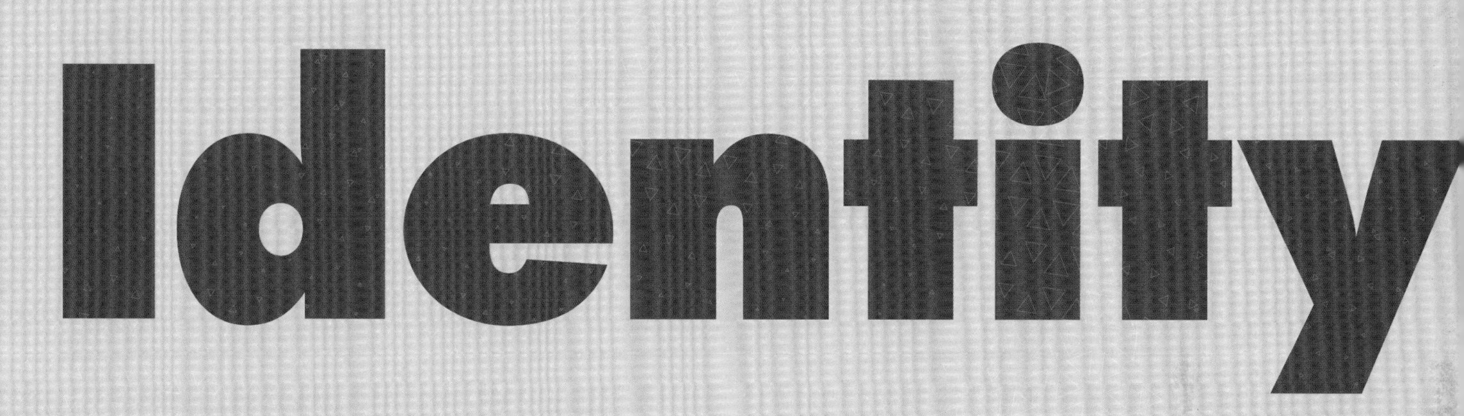

标识与身份的体现

Embodying Brand and Identity

Silvio Carta

作为其市场战略的重要组成部分，每一家公司都会开发出强大而清晰的品牌标识，与客户产生共鸣。这种策略在那些对该公司及其产品感兴趣、被吸引或仅仅注意到该公司的人之间建立了联系。在某些情况下，公司的形象比产品本身更重要，它是建立公司公众形象的基础。通过它的公众形象，公司不仅能够说服人们成为顾客，即购买它的产品，更重要的是，随着时间的推移，还能留住这些顾客，建立对品牌的忠诚和依恋，将之作为一个抽象的实体，而不是特定的产品。公司的品牌推广是通过复杂的机制系统来运作的，涉及日常生活的个人、心理和社会方面。关键因素在公司公众形象的塑造中发挥了作用。其中之一是社会愿望，也就是通过拥有某种产品，一个人可能会在他/她的社会领域中感觉更好，更容易得到认可。另一种强大的心理杠杆可能是，我们每个人都重视自己属于某个社会群体的感觉。换句话说，就是不希望有被社会排斥在某个群体之外的消极感觉。这种产品也就成了一种进入该特定群体的令牌。拥有某一产品或某一特定品牌的任何产品，可能会成为一个人被社会认可为该群体或社群成员（例如，该特定产品的用户）的方式。建立公司形象的另一个方面可能是对产品质量的信赖和信心。在通常的情况下，客户购买是因为他们相信公司的信誉，以及他们所购买的产品品质高、安全可靠。

公司通常可以围绕可持续发展的理念来树立自己的形象，其中对环境的关心、对自然和真正品质的欣赏、对地球和社区的尊重都得到了高度重视（以FairTrade或森林管理委员会的倡议为例）。每家公司所创造的形象都是更广泛战略的一部分，在这个战略中，公司的价值观、工作精神和使命宣言都有助于以一种非常清晰的方式向公众展示公司本身及其产品。

这里所介绍的项目都是很有价值的实例，它们展示了公司如何通过自己的建筑向世界展示他们的公共形象。以下项目可以被划分为两大类：第一类所包含的公司，其产品在很大程度上与公司本身是一致的，因此品牌形象非常强大，得到了广泛的认可，以至于公司名称对大多数人来说就意味着产品。

20世纪30年代，丹麦乐高公司成立，通过游戏、体验和幻想刺激，传递了一种强有力的启发和教育信息。该公司在他们的网站上解释说："在公司精神的指引下：'只有最好的才足够好'，公司致力于儿童的发展，旨在通过富有创造性的游戏和学习来激励和培养明天的建设者。"如果要用几个词来描述乐高玩具，你可能会用"好玩""五颜六色""积木""色彩火花""有创意"等等，而BIG设计的乐高之家（154页）正是如此。体量的概念在这里很突出，没有硬性的组合，从而暗示了组合块体的无限可能性。每个体量之间都有空隙和空间供自然光线穿过或人通过。建筑物的每一个区域看起来都与其他区域不同，但又都是用相同的设计原则设计的。这表明了乐高积木在建造空间和物体时所带来的自由和创造力。一层的特色是一个基于广场的大型开放区域，没有特定的空间层次结构。游客可以从任何地方往任何方向行走，再次强调了这些体块所提供的极高的自由度。也许这个项目比本部分所介绍的其他项目采用了更直接的方式，通过建筑设计直接体现了公司的使命（精神），其中所有的传统物理元素（光线、颜色、组成、体量、材料等）在游客的体验中扮演着重要的角色。乐高之家就是这家丹麦公司多年来一直在全力打造的空间形象。

韩泰轮胎中央研发中心（72页）位于韩国大田的Daeduk Innopolis，由Foster+Partners建筑事务所设计，是建筑设计如何帮助实现公司宗旨的另一个实例。这座96328m²的研发中心旨在"体现创新和技术的领先地位"，"（该公司）在高科技建筑领域的承诺，通过将最新技术和材料融入建筑设计，得到了很好的体现"。韩泰轮胎公司的愿景是通过一座功能强大、令人印象深刻的复杂建筑实现的。该建筑群由一系列的翼形结构和圆形建筑组成，在同一个椭圆形屋顶下。为了营造复杂和光滑的整体感觉，建筑结构内外的曲线在高度和

As an essential part of its market strategy, each company develops a strong and clear brand identity that resonates with its customers. This strategy establishes a connection between those who are interested, allured, or simply notice the company in the market and its products. In some case more than the product itself, the image of the company is fundamental to establish the public presence of the company. Through its public image, the company not only is able to convince people to become customers, namely to buy its products, but more importantly to retain these customers over time, building loyalty and attachment to the brand as an abstract entity (as opposed to a specific product). The company branding works through a system of sophisticated mechanisms that involve the personal, psychological and social aspects of everyday life. Key elements come to play in the creation of the public image of the company. One of them is the social aspiration, whereby, through the ownership of a certain product, a person may feel better and more established in his/her social sphere. Another powerful psychological leverage can be the importance that each of us gives to the sense of belonging to a certain social group. Put it in other terms, the negative feeling of being socially excluded from a certain group of which one wants to be part. The product becomes an entry token to that particular group. The fact of owning a certain product, or any product from that particular brand may become the way an individual is socially recognised as a member of that group or community (e.g. the users of that particular product). Another aspect around which a company image is built can be the trust and the faith in the quality of the product. Often customers buy because they believe in the respectability of the company and in the fact that the product they are acquiring is of high quality, safe, and reliable. Companies can often build their image around a sustainable idea, where the care for the environment, the appreciation for natural and genuine components, the respect for the planet and communities are highly considered (think of the FairTrade or the Forest Stewardship Council initiatives as an example). The image that each company creates is part of a wider strategy, where the company values, the working ethos and the mission statement contribute to the same goal of presenting the company itself and its products to the public in a very clear manner.

The projects presented in this section are valuable examples of how companies translate their public images to the world through their own buildings. The following projects can be categorised into two main groups.

The first one includes companies of which products are largely identified with the company itself, whereby the branding is so strong and widely recognised that the company name means the product for the majority of

长度上与宏大的空间交替。内部庭院和翼形结构之间的巨大空隙以未来主义的空间为特征，它们带有光滑的反光覆层和在挑空空间中悬浮的会议区域。韩泰轮胎中央研发中心的建筑设计体现了该公司的承诺，他们致力于通过研究和开发，在未来的技术领域占据全球领先地位。这座新建筑体现了所有这些价值，为未来的研究和发现提供了实体空间。

乐高公司和韩泰轮胎销售的是实实在在的产品，而西班牙对外银行新总部（52页）则提供的是金融服务。位于西班牙马德里的西班牙对外银行新总部由赫尔佐格和德梅隆建筑事务所设计，它集中体现了该公司的意图和价值观。现任总裁兼集团执行主席弗朗西斯科·冈萨雷斯解释说，该银行的宗旨是"为每个人带来一个充满机遇的时代。西班牙对外银行的团队是基于诚信、审慎和透明原则的企业文化的一部分，这些原则对我们建立强大的商业模式至关重要，这种商业模式能够成长，而且以客户为中心"。他们的新总部由三层建筑组成，由一组桥梁、花园、庭院和通道连接。建筑师决定创造一座内向型的建筑，雇主可以在建筑之间享受高品质的空间，让合作变得更加容易。"建筑的层数不高，这种安排促进了交流：人们不再乘坐电梯，而是走上楼梯，鼓励了非正式的交流；最大化的视觉透明度给每个人都提供了一个视角，产生一种社区感；而相对较小的单元让员工更容易认同他们专属的工作组。"建筑师解释道。总部的这些内部街道都成了人们见面、讨论，以及在办公室和会议之间移动的公共空间。在这些公共区域散步时，一切都是透明的，同时又受到保护，给人一种安全感和谨慎感。尽管低层建筑和通道、街道和花园系统保证了雇主和访客之间流畅、持续的视觉和身体交流，但这座93m高的大楼还是让西班牙对外银行总部成为西班牙首都的地标性建筑，从远处就醒目地表明了其存在。

people.
The 1930s founded Danish company LEGO promotes a strong message of inspiration and education through play, experience and fantasy stimulation. On their website, they explain that "guided by the company spirit: 'Only the best is good enough', the company is committed to the development of children and aims to inspire and develop the builders of tomorrow through creative play and learning". If asked to describe LEGO with a few words, one might possibly use "playful", "colourful", blocks", "sparks of colours", "creative", etc. BIG's LEGO House (p.154) is exactly this. The idea of volumes is here prominent, without a rigid composition so as to suggest infinite possibilities of combining the blocks. There are gaps and spaces in-between each volume for natural light or people to pass through. Every area of the building appears different from the others yet built by using the same principles. This suggests the freedom and creativity that the LEGO blocks allow while building spaces and objects. The ground floor is characterised by a large open area based on a square, and with no particular spatial hierarchy. Visitors can go in any direction from anywhere, again to emphasise the high degree of freedom that the blocks offer. Perhaps in a more direct way than the others presented in this section, this project is a direct materialisation of the company mission (spirit) through architecture, where all its traditional physical elements (light, colour, composition, volumes, materials, etc.) play a fundamental part in the visitors' experience. The LEGO House is the image that the Danish company built over the years making spatial at full scale.
Located in Daeduk Innopolis, Daejeon Korea and designed by Foster + Partners, the Hankook Technodome (p.72) is another example of how architecture can help to materialise a company ambition. This 96,328m^2 research and development center has been designed to "reflect innovation and technology leadership" where "[the company's] commitment in high-tech architecture, is well represented by incorporating the latest technology and materials with architecture design". Hankook Tire's vision is materialised through a sophisticated, functional and impressive building. This compound consists of a series of wings and circular buildings under the same oval roof. In order to contribute to the overall feeling of sophistication and slickness, sinuous curves on the inside and outside alternate with grand spaces, both in height and in length. The inner courtyards and the big voids in-between the wings are characterised by futuristic spaces, with smooth and reflective cladding and meeting areas suspended in the voids. The architecture of the Hankook Technodome is the expression of the company's commitment to global leader-

第二类包括提供多种产品和服务的大公司和企业集团，这些公司的名字通常还没有它们的产品和服务有名。在这种情况下，公司的价值观和整体形象需要在构建时考虑到这种多样性。这些公司的形象需要传达一些关于他们企业的合作性和包容性的性质。

例如，韩国美容化妆品集团爱茉莉太平洋"致力于成为亚洲之美创造者"，通过30个涵盖美容、家居和保健产品的品牌，为全球客户提供了一整套的美容和健康解决方案。由大卫·奇普菲尔德建筑师事务所设计的新总部（110页）可以看作是该公司公众形象的最终表达。这座建筑试图融入周边密集的城市环境，然而，更近距离的观察揭示了它独特的微妙之处。首先是材料和光线的应用。整座建筑都覆盖着透明的百叶窗系统，赋予建筑充满活力的外观。通过关注该建筑的设计细节，人们能够欣赏到立面的质感和丰富性，否则看到的只是统一的外观而已。夜晚，人造灯光揭示了不同建筑体块的复杂性，让行人猜测这些新总部大楼都有哪几项活动和功能，让人们更好地欣赏建筑的立面元素和内部的空间性。第二个是体量，它看起来像一个规则的固体。随着更多的关注，人们可能会注意到，与室内花园和建筑不同块相对应的立面是略微突出的。"建筑的形式是抽象的，也是有自己的姿态的。建筑围绕着一个中央庭院精心设计，集中在一个清晰的单体体量上，从而最大化自然通风和日光的效果。"建筑师解释道。这是需要观察者通过关注细节、比例和细微差别，去仔细、缓慢欣赏的建筑。它清楚地表明了该公司的本质：其化妆品和美容产品均为公司最重要的业务。建筑师是这样描述的："打个比方，新建筑呼应了现代组织的愿望，在本地和全球、私人和公共、集体和个人、正式和非正式之间进行协调，而这样做是为了建立其动态特性。"

ship in future technology through the use of research and development. This new building embodies all these values, providing a physical place for future research and discoveries.

Whereas LEGO and Hankook sell a concrete and tangible product, the New Headquarters for BBVA (p.52) offers financial services. Designed by Herzog & de Meuron, their new headquarters in Madrid, Spain epitomises the business' intentions and values. Current president and group executive chairman Francisco González explains that the bank's purpose is "to bring the age of opportunity to everyone. The BBVA team are part of a corporate culture based on the principles of integrity, prudence and transparency, which have been key in allowing us to build a strong business model that is able to grow and is very customer centric". Their new headquarters consist of a compound of three storey buildings linked by a system of bridges, gardens, courtyards and passages. The architects decided to create a building that looks inwards, where employers can enjoy high-quality spaces in-between buildings and where collaboration becomes easy. "The low-rise arrangement fosters communication: instead of taking elevators, people walk up stairs that encourage informal exchange; maximized visual transparency gives everybody a view and generates a sense of community; whilst the relatively small units permit employees to identify with their particular workgroup." explain the architects. The inner streets of these headquarters become a public space for people to meet, discuss and move between offices and meetings. While walking in these public areas everything is transparent, yet protected, conferring a sense of security and prudence. Whereas the low-rise buildings and the system of passages, streets and gardens guarantee a fluid and continuous visuals and physical communication for the employers and visitors, the 93 meters tower confers the BBVA headquarters with a landmark in the Spanish capital, signalling their presence from distance.

The second group includes large companies and conglomerates that offer a diverse range of products and services, where the company's name is often widely less known to the general public than their offer. In this case, the company's values and overall image need to be built with this diversity in mind. The image for these companies needs to communicate something about the corporative and inclusive nature of their businesses.

The Amorepacific, for example, is a South Korean beauty and cosmetics conglomerate which is "committed to becoming the 'Asian Beauty Creator' and has provided global customers with a total package of beauty and health solutions through 30 brands spanning beauty, household, and healthcare products". Designed by David Chipper-

Natura是一家巴西化妆品及美容产品跨国公司。"他们的产品很好地表达了公司的本质：公司为了开发新产品，雇用了一批人，这些人能够将科学知识与对巴西丰富多样化植物的可持续利用相结合"（作者的翻译）。这座新总部（170页）由圣保罗的Dal Pian Arquitetos建筑师事务所设计，紧邻他们的配送中心，表现了该公司对于可持续生物多样性的追求，以及对当地价值和科学研究的关注。这个项目的特点是具有不同层次的花园，在这里，室内和室外都栽种了绿色植物。巨大的中央空间不但方便使用者自由行走，还有利于自然光线入射和通风。该建筑围绕可持续建设原则进行设计，在不同的立面上设置不同类型的百叶窗，以控制太阳得热、直接光照和总热量。绿色屋顶和材料的选择有助于建筑整体的高水平可持续发展。同样，建筑师还通过复杂的建筑外围护结构和各种过渡空间来表达设计的美观和对细节的关注。

默克公司是一家专注于医疗、生命科学和性能材料的全球化大公司，于1668年在德国达姆施塔特成立。默克公司所追求的价值观是创新、变革，以及突破性发现的重要性和卓越创意的力量。"它们在改变着我们的生活，影响着我们应对重大疾病的方式，是我们赖以生存的物品不可或缺的一部分。从癌症治疗和实验室工具，到智能手机屏幕或化妆品，科学和技术无处不在，以不同的方式影响着我们的生活。"该公司的主要目标是"改变数百万人的生活"。他们在达姆施塔特的新总部（188页）由HENN建筑事务所设计，被设想为一个创新的空间，以培养未来在科学和技术方面的卓越理念。建筑的内部空间是流动而连续的，在这里，各个工作组或区域通过桥梁、坡道、台阶或仅仅通过开放空间进行视觉连接。特别值得一提的是，中央主要区域的特点是设计了一套复杂、流畅、舒展的桥梁和

field Architects, their new headquarters (p.110) can be seen as the ultimate expression of the public image of this company. The building tries to blend in its dense urban context, yet a closer view reveals the existence of subtleties that make it unique. The first is the materiality and the use of light. The entire building is clad with a system of diaphanous brise-soleil that confers the building a vibrant and dynamic look. By focusing on the details of this building one is able to appreciate the quality and the richness of the facades, which would otherwise seem simply uniform. At night, the artificial lights let appreciate even better the facades elements and the inner spatiality of the building, by unveiling the complexity of the different blocks and let passers-by guess on the several activities and functions that these new headquarters house. The second is the volume which appears like a solid and regular block. With more attention, one may notice that the facades in correspondence of the interior gardens and different blocks of the building are slightly protruding. "The form of the building is both abstract and gestural. Focusing on a single, clear volume, the proportions of the building have been carefully developed around a central courtyard to maximize the effectiveness of natural ventilation and daylight." explain the architects. This is a building that the observer needs to appreciate carefully and slowly, by focusing on details, proportions and nuances. This clearly suggests the very nature of the company, whereby cosmetics products and beauty are at the forefront of the business. The architects described that: "Metaphorically, the new building echoes the aspirations of a modern organisation, mediating between local and global, private and public, collective and individual, formal and informal, and in doing so establishing its dynamic identity."
Natura is a Brazilian multinational company of cosmetics and beauty products. "Their products are the best expression of their essence: in order to develop them, they employ a network of people that are able to integrate scientific knowledge with the sustainable use of the rich Brazilian botanical biodiversity" (translation of the author). Designed by São Paulo-based Dal Pian Arquitetos and built next to their distribution center, this new Headquarters (p.170) is the expression of the company quest for sustainable biodiversity, attention for local values and scientific research. This project is characterised by a number of gardens at different levels, where the greenery is present both on the inside and the outside. A large central void allows for free circulation of people, natural light and ventilation. The building has been designed around sustainable construction principles, with different types of louvres in the different facades to control the solar gain, direct light exposure, and overall heat. The green roof and the

坡道系统，该系统连接了不同的楼层和原有的对角线景观，以及为用户提供的连续交通路线。"创新是由好奇心推动的。在达姆施塔特的总部，我们创造了一个完美的环境来培养这两种人才。不过，默克创新中心不仅仅是一座建筑，它还是一种思维模式。"

总部位于首尔的可隆集团是一家多元化的公司，其业务范围从纺织品、化学品和可持续技术，到运动和时装成衣市场的原创服装系列，包含甚广。该公司不仅生产特定的产品，更具有明确的使命和整体公众形象，致力于研发未来的解决方案。"随着全球竞争的加剧，研发竞争力变得更加重要。[…]可隆公司及研发总部（90页）通过不断创造未来的增长动力，提供先进的材料和技术，将发挥核心作用，使公司一跃成为全球领先的特种化学品和材料公司。" 这家公司雄心勃勃的声明完美地反映在他们的新大楼上。该项目由Thom Mayne/Morphosis建筑事务所按照公司的合作创新理念而设计。"该建筑将研究人员、领导和设计师聚集在同一个地方，它结合了灵活的实验室设施、行政管理办公室和活跃的社交空间，鼓励人们在整个公司内部进行更大的互动和交流"。建筑师解释说。这个综合体由五个主要元素组成：总部、实验室和中试实验室、社交空间和公园。尤其是错综复杂的西立面为可隆公司及研发总部打造了一个强大的公共形象。西侧立面由内衬系统和模块化玻璃立面组成，其特点是使用玻璃纤维增强聚合物（GFRP）元素作为遮阳材料，同时赋予中庭（社交空间）引人注目的氛围。总长度为100m、高度为30m的社会空间是新研究园区的重要组成部分。参观者和研究人员可以

choice of materials contribute to the overall high level of sustainable aspects of this building. Again, beauty and attention to details are communicated through a sophisticated building envelop and articulated spaces.

The values pursued by Merck, the global healthcare, life science and performance materials company founded in Darmstadt, Germany in 1668 are innovation, change, and the importance of ground-breaking discoveries and the power of brilliant ideas. "They are life-changing, influencing how we tackle major illness, and integral to the objects we rely on. From cancer therapies and laboratory tools, to screens on smartphones or make-up – science and technology are everywhere and touch our lives in diverse ways." The company's main goal is "to make a difference to millions of lives". Their new headquarters in Darmstadt (p.188), designed by HENN has been conceived as an innovative space to foster the brilliant ideas of the future in technology and science. The interior space of the building is fluid and continuous, where the individual working groups or areas are connected through bridges, ramps, steps, or simply visually through the open void spaces. In particular, the central main area is characterised by a complex, yet fluid system of bridges and ramps that unfolds, connecting the different floors and originating diagonal views and continuous circulation for the users of the building. "Innovation is driven by curiosity. At our Darmstadt headquarters we have created the perfect surrounding to foster both. Merck Innovation Center is not just a building though, it's a mindset."

The Kolon Group, based in Seoul, is a diverse corporation of which activities range from textiles, chemicals, and sustainable technologies, to original clothing lines in the athletic and ready-to-wear fashion markets. More than one specific product, this corporation is more identifiable with its mission and overall public image dedicated to research and development and future solutions. "R&D Competitiveness becomes more essential because of the increased global competition. […] By creating constant future growth power and supplying advanced materials and technologies, Kolon Corporate and Research Headquarters (p.90) will play a core role in order for KOLON Industries, Inc. to make a vaulting jump to become Global Top Company of specialty chemicals and materials". The ambitious company statement is perfectly reflected on their brand-new building. Designed by Thom Mayne/Morphosis, this project revolves around the company idea of being innovative through collaboration. "Bringing researchers, leadership, and designers together in one location, the building combines flexible laboratory facilities with execu-

对每一个楼层都有全面的了解，还可以透过精心设计的立面系统欣赏到室外首尔的秀丽风光，这正是该公司创新使命的不断展示。Morphosis建筑事务所通过尖端、大胆的建筑设计，为可隆中央研究公园的雄心壮志和未来思维提供了最佳外形和形式。

Kiswire公司是一家全球性的钢铁制造商，总部位于韩国，业务范围从为汽车行业生产高质量的电线，到为钢琴琴弦生产琴用钢丝，范围很广。该公司的简介描述了他们所提供产品的广阔范围，包括钢琴弦、金属桥、汽车和电梯；"远比你想象的要多得多，Kiswire近在眼前，让世界变得更美好"。Kiswire的公众形象强调了其产品的可靠性、强度、耐用性和灵活性。该项目由韩国的BCHO建筑师事务所设计，反映了所有这些雄心壮志。他们的新总部（130页）是一个复杂的综合体，花园、建筑体量、亭台楼阁、绿色屋顶、展览空间、桥梁、花园和雨篷都被结合在一个大型项目中。这个综合体的名字叫F1963（F代表工厂，后面跟着公司成立的年份）。F1963"是一个自然与艺术共存的地方，一年365天，各种类型的文化和艺术活动在这里都能找到。F1963的愿景是成为釜山当地社区的文化中心"（公司网站）。BCHO的设计展示了建筑元素在同一个综合体中的多种运用方式，打造出灵活、适应性强而又令人惊讶的空间。F1963是源于工业聚落的多元文化综合体，它证明了公司的愿景是在一个"创造和消费同时发生的场所（创造、展示、消费），不断提高现有的生产、工作和生活质量"。这个新项目实现了Kiswire公司的多用途产品的理念，这些具有相同品质和优势的产品往往具有不同的用途。

tive offices and active social spaces that encourage greater interaction and exchange across the company." explain the architects. This compound consists of five main elements: the headquarters, the labs and pilot labs, the social space and the park. In particular, the sophisticated west facade provides the KCRP with a strong public image about the Kolon Group. With a system of interior liners, a modular glass facade, the west elevation is characterised by glass fiber reinforced polymer (GFRP) elements that act as sunshades, as well as conferring the atrium (social space) a dramatic atmospheric quality. With a total length of 100 meters and a height of 30, the social space is a key part of the new research park. Visitors and researchers can have a total overview of each floor and can see the outside view of Seoul mediated by the elaborated facade system, as a constant demonstration of the company's innovative mission. Morphosis spectacular architecture gives probably the best shape and form to the ambitions and future thinking of KOLON Central Research Park through a cutting edge and bold architecture.
Kiswire is a global Korea-based steel manufacturer of which business ranges from producing high-quality wires for the car industry, to music wires for piano strings. The company's profile describes the breadth of the products they offer, referring to piano strings and metal bridges, to cars and lifts; "much more than you realise, Kiswire is near, making the world a better place". Kiswire's public image is all about reliance, strength, durability and the flexibility of use that characterises their products. The project designed by the Seoul-based BCHO Architects reflects all these ambitions. Their new headquarters (p.130) are a complex compound where gardens, building blocks, pavilions, green roofs, exhibition spaces, bridges, gardens and canopies are combined in one large project. This complex goes under the name of F1963 (where F stands for Factory, followed by the year of establishment of the company). F1963 "is a place where nature and art coexist, and all genres of culture and arts are available 365 days a year. F1963's vision is to become a cultural hub for the local communities of Busan" (company website). BCHO's design shows the variety of use of architectural elements in the same compound, where space becomes flexible, adaptable and surprising. F1963 is multicultural complex that originates from an industrial settlement. This is a testimony to the company's vision of continuously improving the existing quality of production, working and living in a "place where creation and consumption take place (Creating, Showing, Consuming)". This new project materialises Kiswire's idea of versatile products that can offer always different uses with an equal level of quality and strength.

西班牙对外银行新总部

New Headquarters for BBVA

Herzog & de Meuron

BBVA
BBVA

首都周边

西班牙对外银行的新总部位于马德里北部边缘。场地面对着高速公路，周围是新建的办公楼、商业建筑和住宅开发项目。当银行收购这块土地时，八幢未完工的办公大楼占据了相当大的一块土地，而尽可能多的现有建筑将被纳入新的开发项目。

一块地毯

带有庭院、通道和灌溉花园的线性结构三层建筑覆盖了整个场地，场地的坡度相当大，就像地毯一样，类似于阿拉伯花园。考虑到场地周围的环境没有太多的特性，没有太多的可关联之处，赫尔佐格和德梅隆建筑事务所选择"内化"这个综合体，围绕西班牙对外银行的内部需求进行独特的设计。

建筑低层的安排方式促进了交流：人们不再乘坐电梯，而是走上楼梯，这种设计鼓励了非正式的交流；最大化的视觉透明度让每个人都有自己的视角，产生了一种社区感；而相对较小的办公单元让员工能够认清他们特定的工作组在什么位置。

新总部的设计规模为6000平方英尺（约合557.4m²）。场地和开发项目的规模要求我们必须寻找一个彻底的解决方案，我们选择在这个原本默默无闻的城市景观中创造一个内部的绿洲，这是一个在大自然和建筑之间建立平衡的地方，就像一座小城市和一个大花园。

建筑师对现有的建筑进行改造，以适应新的结构，也为了能创建类似线性外形和规模的办公室和花园。它们要么被切割出来，要么被结合到建筑结构中。

南方风格的建筑

它是一种原始的建筑设计风格，其中的结构被突出地表达出来。这种设计受到强烈的日照条件的影响，最终形成了一种南方风格的建筑。沿着相当狭窄的内部花园和街道，混凝土柱和悬挑楼板可提供遮阳，以防止过多的阳光入射，这样可以减少对空调的需求。全高的嵌入式玻璃为办公室提供了良好的采光条件，从而最大限度地减少了人工照明。

沿着综合体的外围，我们开发了固定在楼板之间的百叶窗。与著名的现代建筑案例不同，这些百叶窗在较低的部分被切割成一个角度，就是为了在最不需要保护的地方提供更多的视野和日光，从而产生了在方向和大小上随着太阳的角度和房间功能而变化的象征性元素。随着百叶窗高度的调整，倾斜的场地在立面上创造了另一种微妙而有影响力的结果。

将西班牙对外银行嵌入马德里的天际线

一个圆形的广场从地毯上被裁剪下来，然后，它就像一个倾斜向上的物体，变成了一座纤细高大的塔，标志着西班牙对外银行在马德里天际线中的轮廓。与低层办公室形成对比的是，塔楼提供了另一种

©BBVA

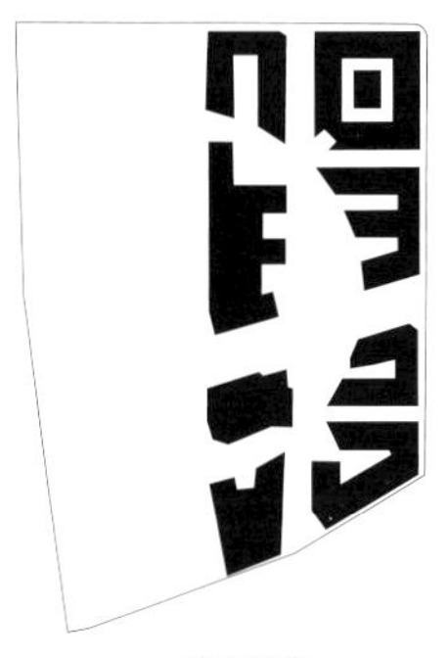

现有建筑
existing buildings

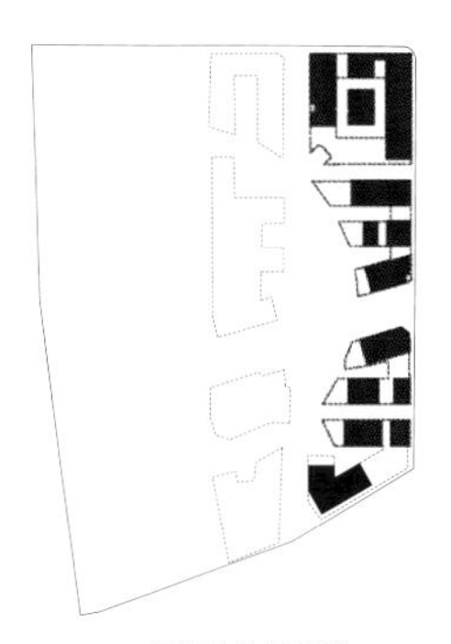

切割结构的拆除
demolition and cut-out structure

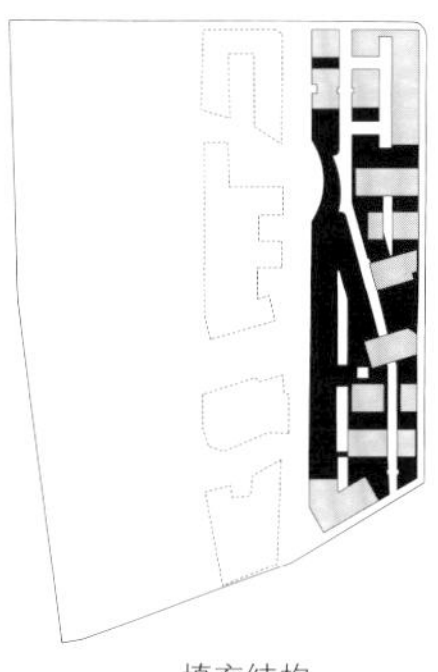

填充结构
filled-in structure

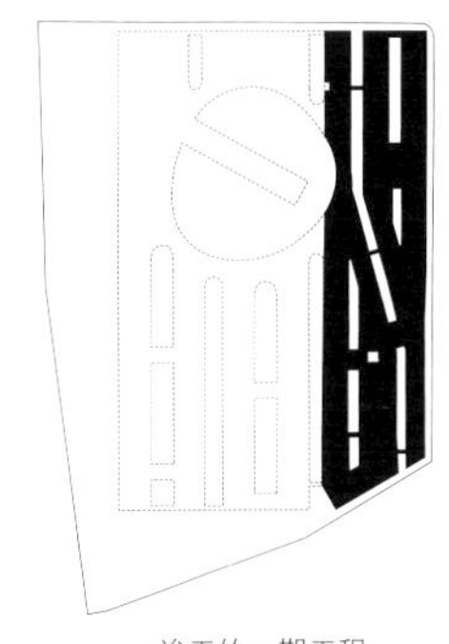

竣工的一期工程
completed phase 1

竣工的建筑（一期+二期）
completed building (phase 1 + phase 2)

类型的工作空间，人们可以欣赏到城市和山脉的景色。广场上种植着数百棵树，周围环绕着各种公共设施。广场和塔楼一起为整个综合体提供了方向。

At the Periphery of the Capital

BBVA's new headquarters is located on the northern periphery of Madrid. The site faces the highway and is surrounded by newly built offices, commercial buildings, and residential developments. When the bank acquired the site, eight unfinished office buildings occupied a substantial portion of the land, and as much of the existing buildings as possible were to be incorporated into the new development.

A Carpet

A linear structure of three-story buildings with courtyards, passages and irrigated gardens is laid over the entire site – which has a considerable slope – like a carpet, analogous to an Arabian garden. Herzog & de Meuron chose to "internalize" the complex, to design it uniquely around the inner needs of BBVA, given that the surroundings didn't have much identity, there simply wasn't much to relate to. The low-rise arrangement fosters communication: instead of taking elevators, people walk up stairs that encourage informal exchange; maximized visual transparency gives everybody a view and generates a sense of community; whilst the relatively small units permit employees to identify with their particular workgroup.

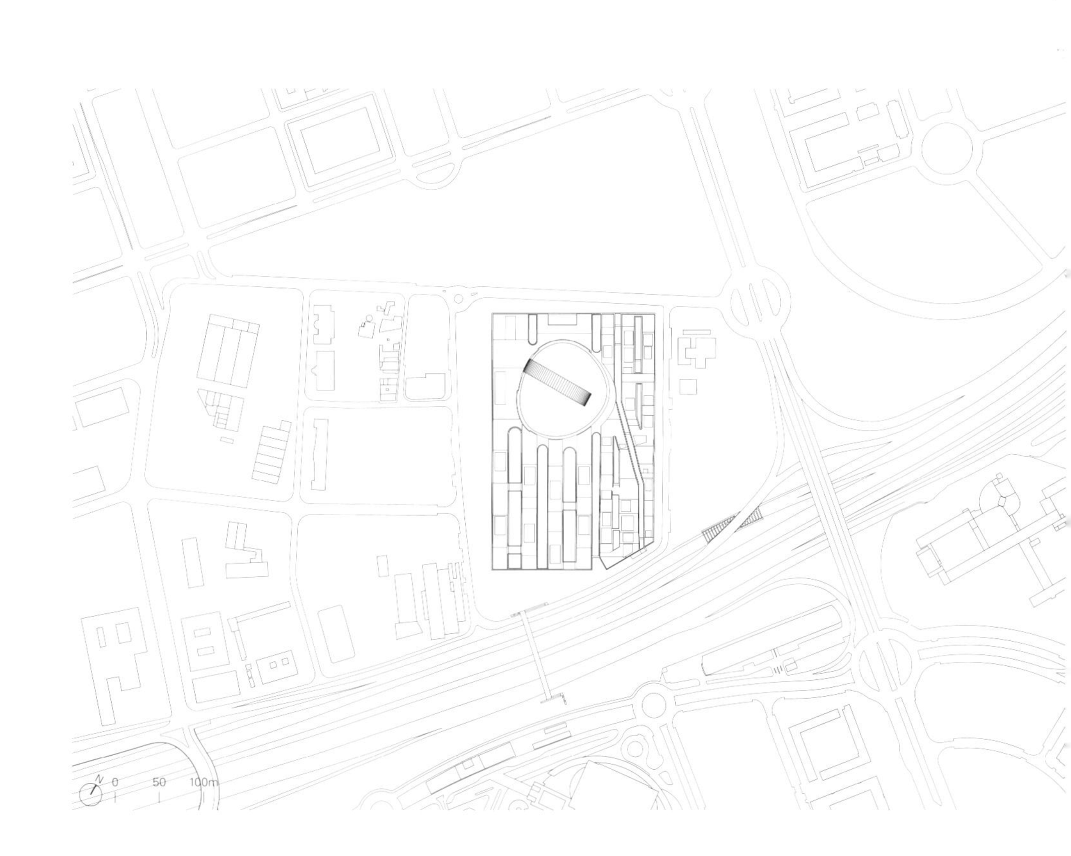

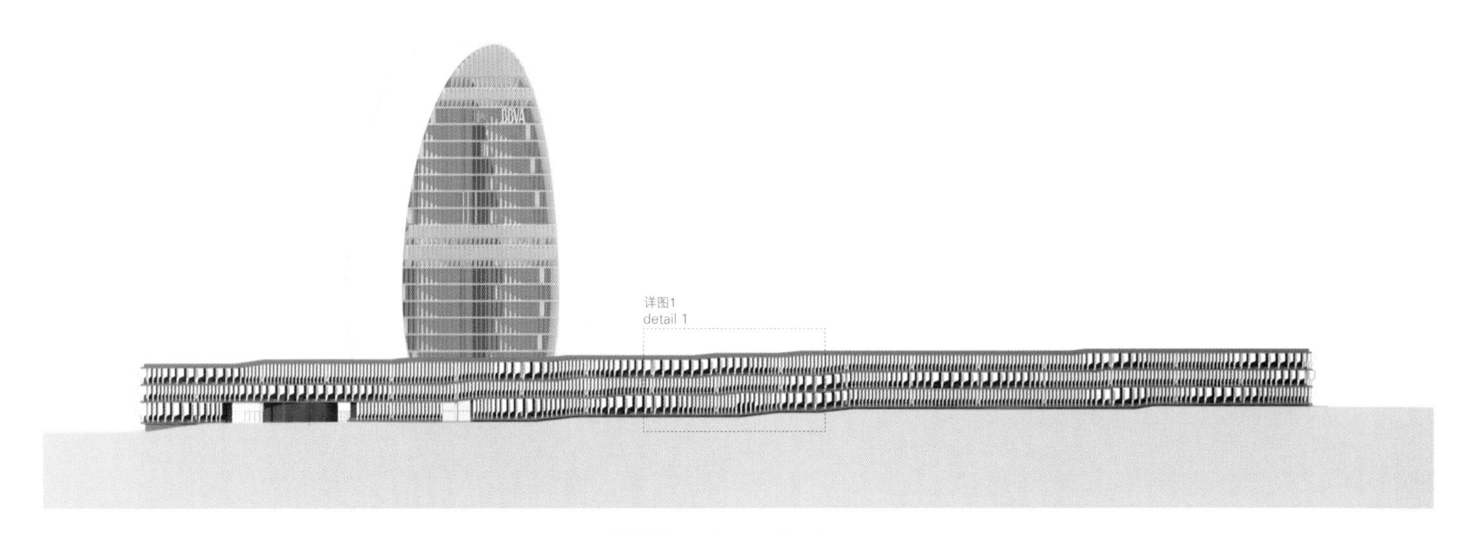

西南立面 south-west elevation

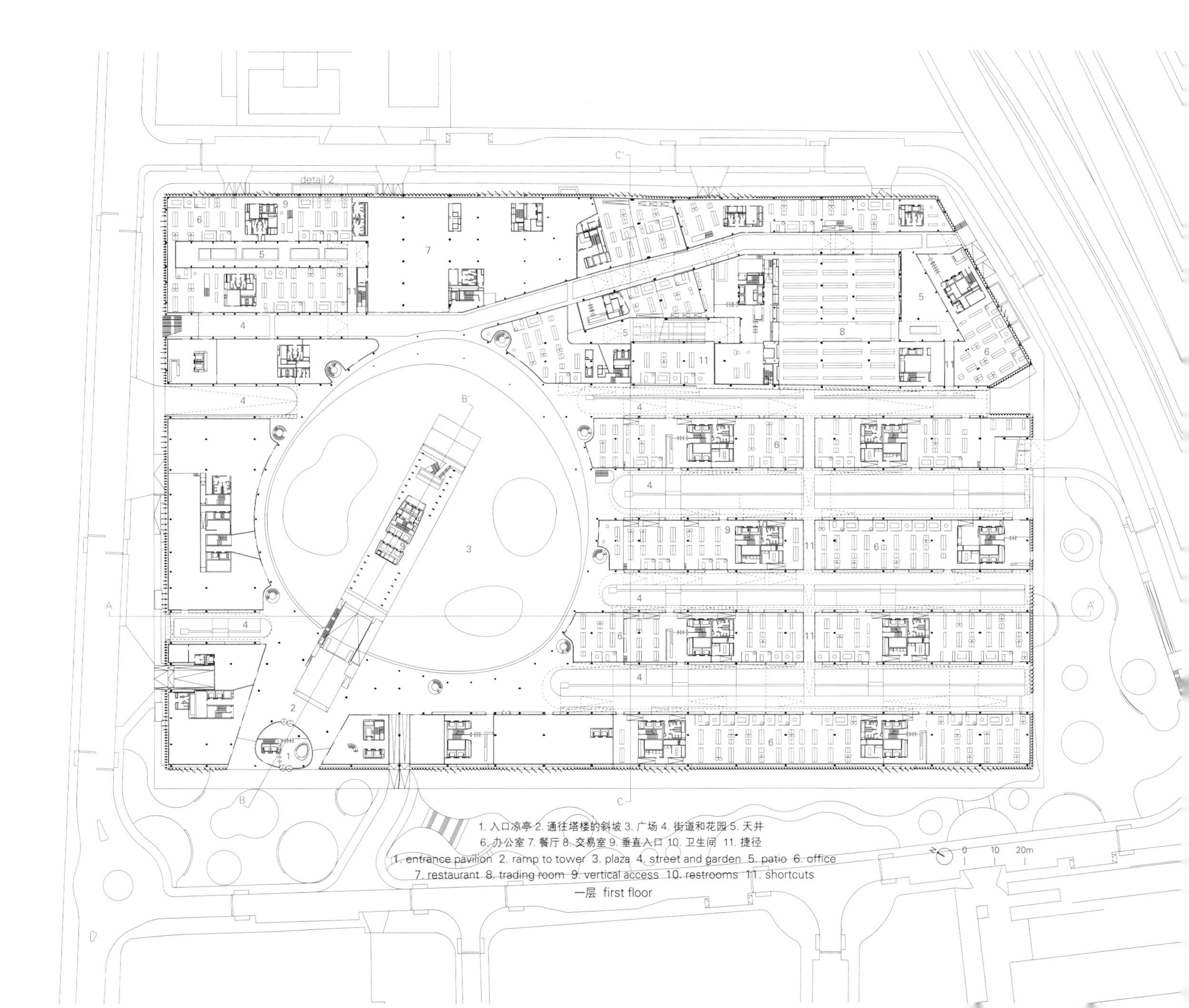

1. 入口凉亭 2. 通往塔楼的斜坡 3. 广场 4. 街道和花园 5. 天井
6. 办公室 7. 餐厅 8. 交易室 9. 垂直入口 10. 卫生间 11. 捷径
1. entrance pavilion 2. ramp to tower 3. plaza 4. street and garden 5. patio 6. office
7. restaurant 8. trading room 9. vertical access 10. restrooms 11. shortcuts
一层 first floor

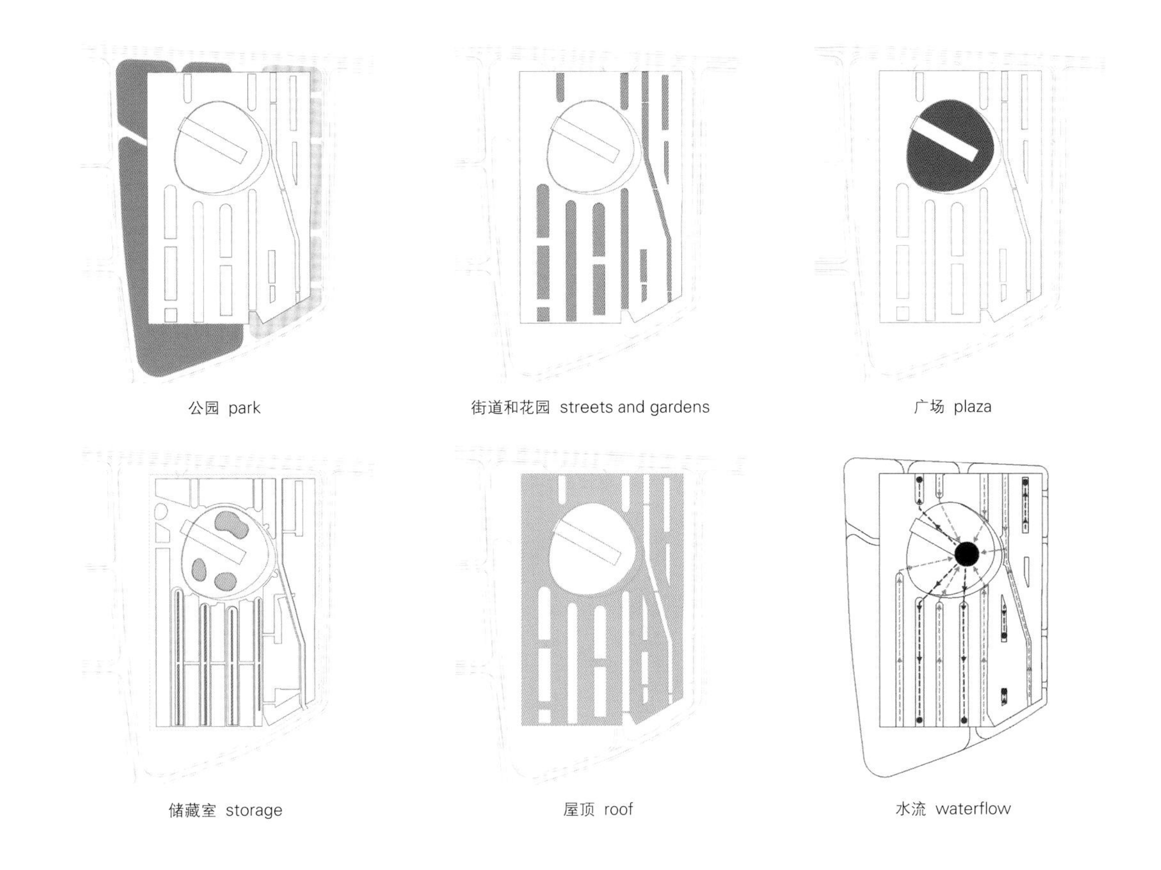

公园 park

街道和花园 streets and gardens

广场 plaza

储藏室 storage

屋顶 roof

水流 waterflow

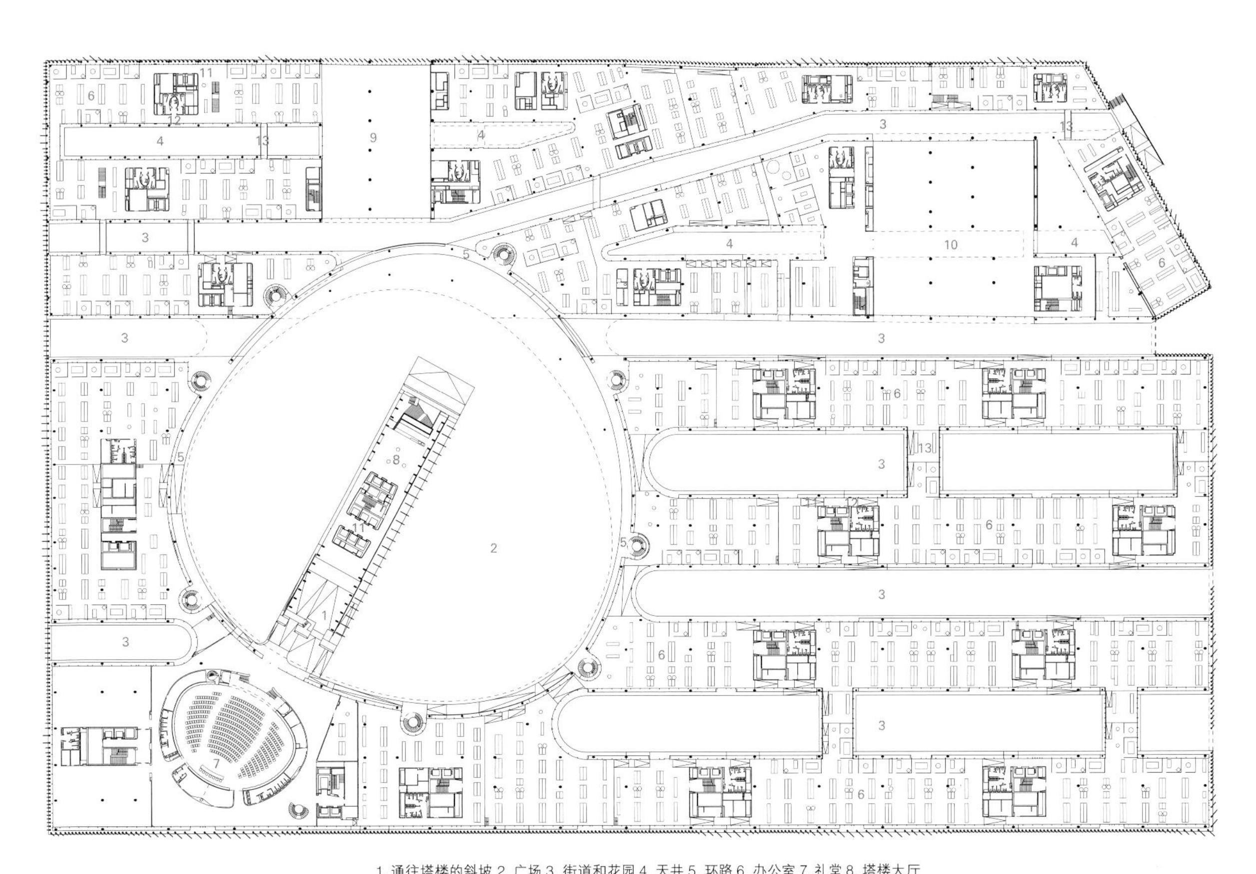

1. 通往塔楼的斜坡 2. 广场 3. 街道和花园 4. 天井 5. 环路 6. 办公室 7. 礼堂 8. 塔楼大厅
9. 餐厅 10. 交易室 11. 垂直入口 12. 卫生间 13. 连桥
1. ramp to tower 2. plaza 3. street and garden 4. patio 5. ring 6. office 7. auditorium 8. tower lobby
9. restaurant 10. trading room 11. vertical access 12. restrooms 13. bridge
二层 second floor

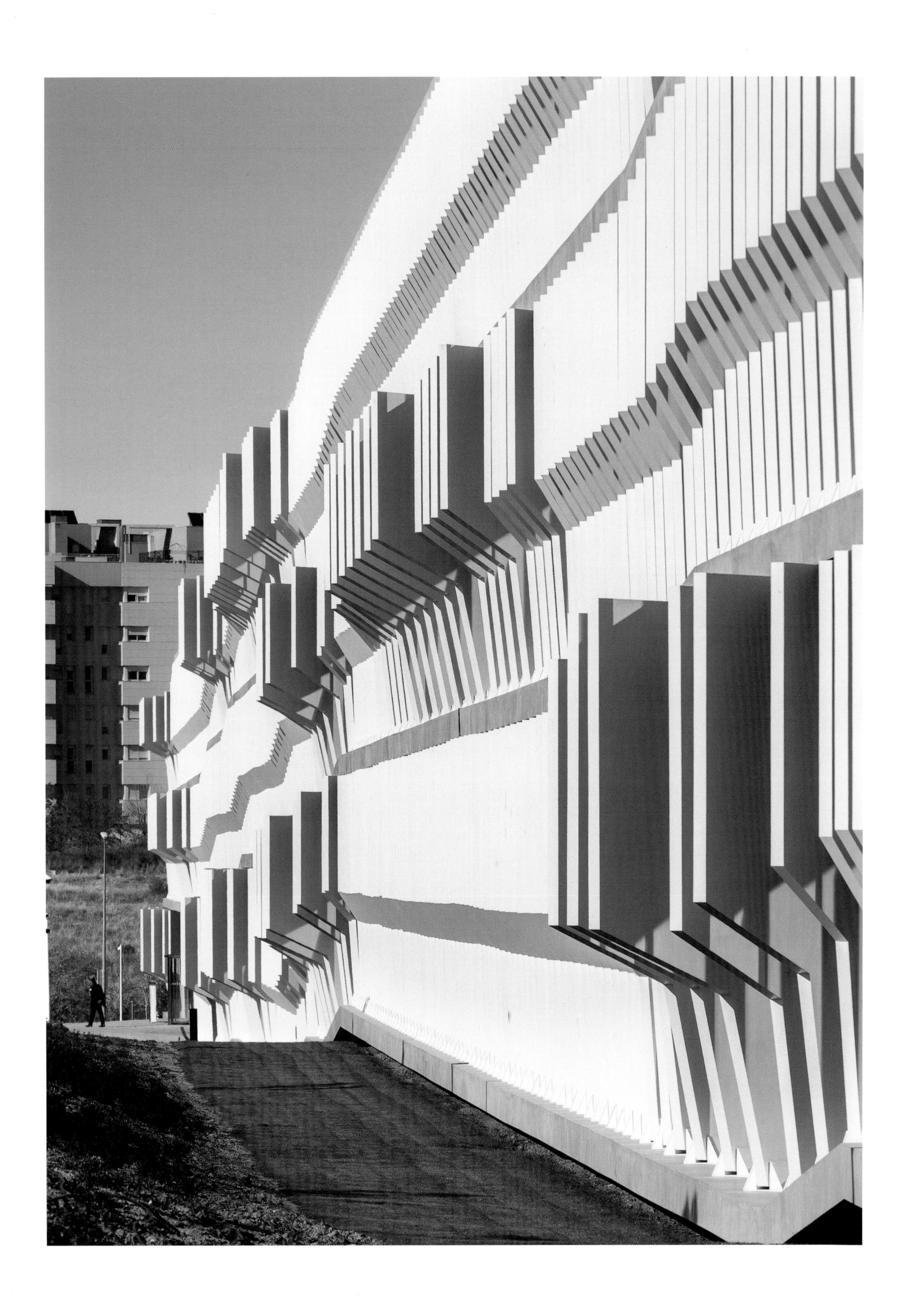

BBVA

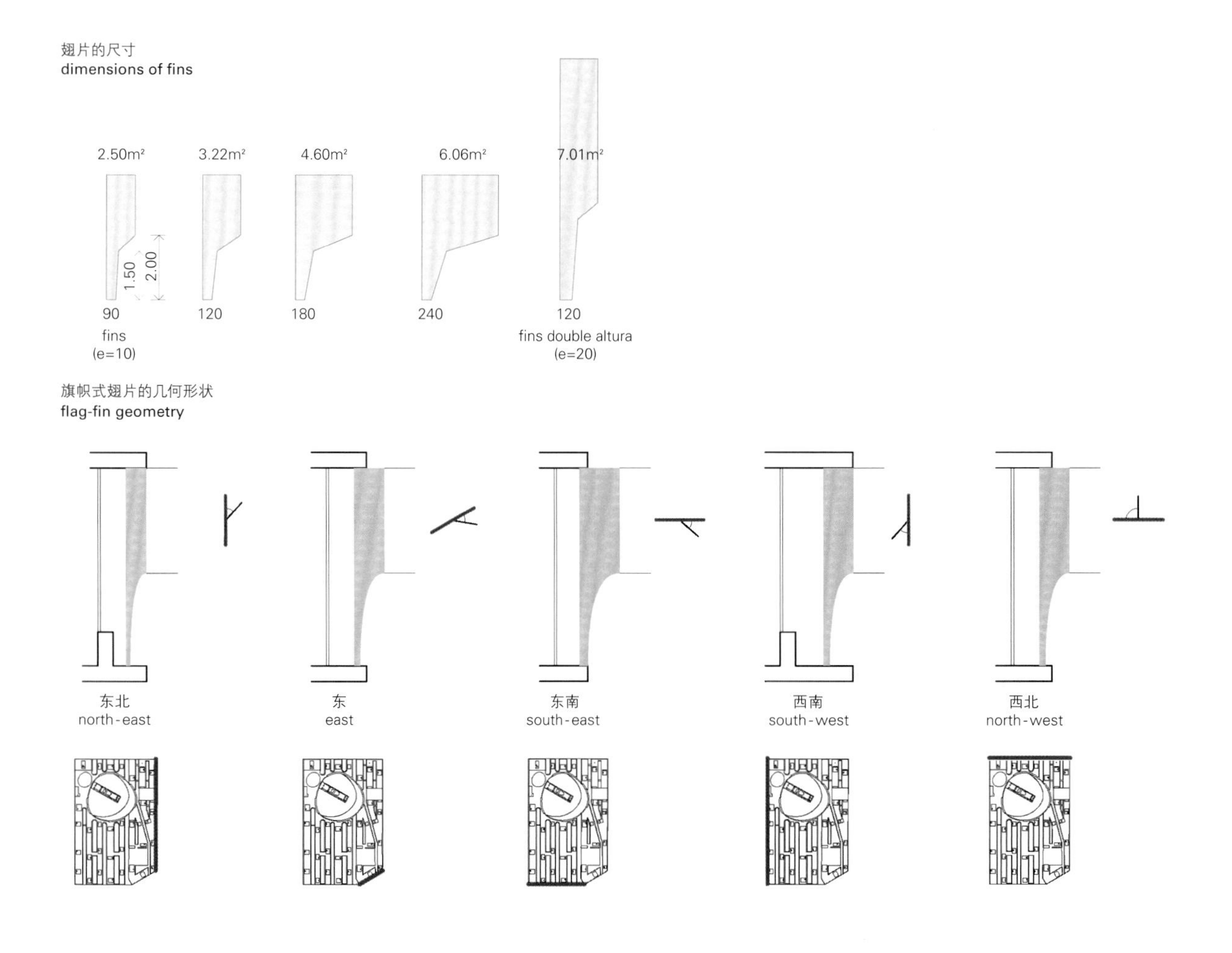

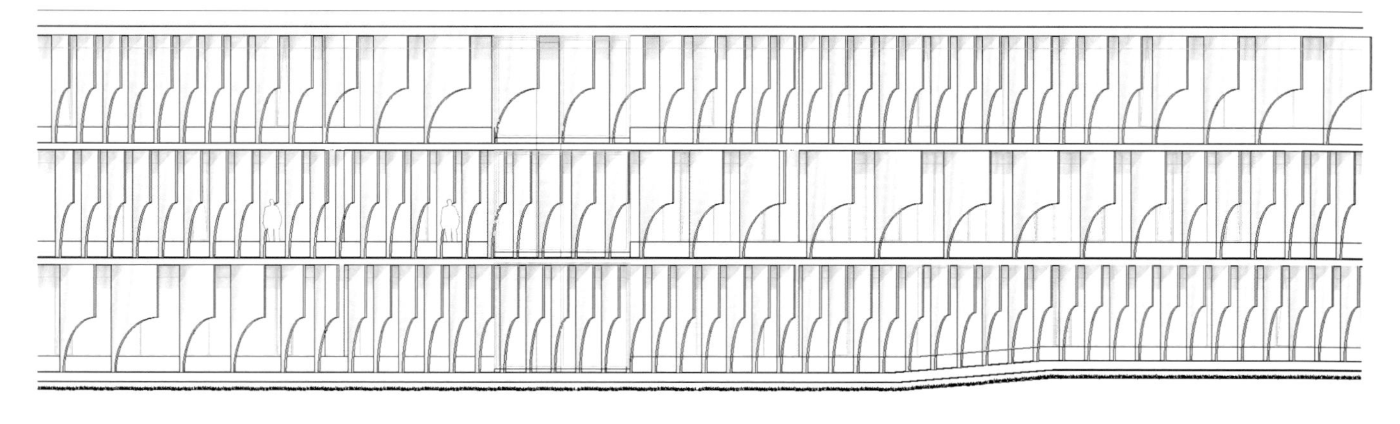

详图1 detail 1

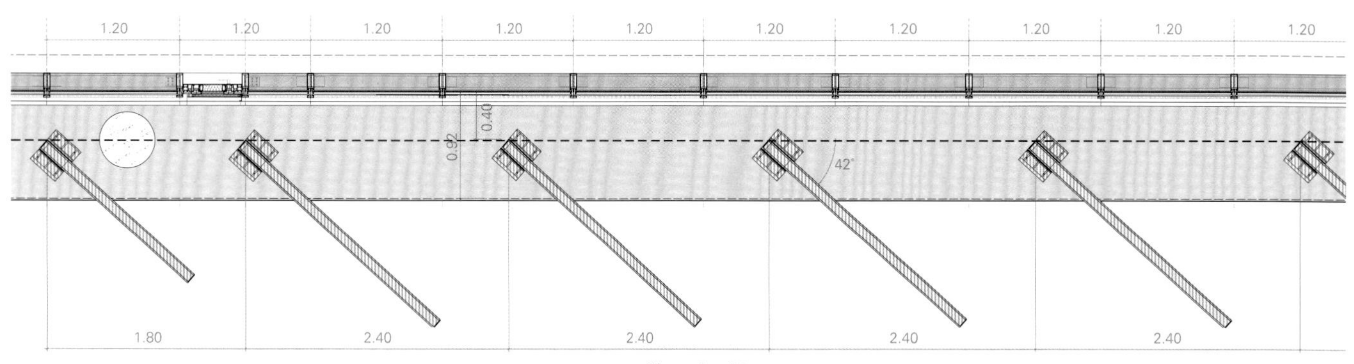

详图2 detail 2

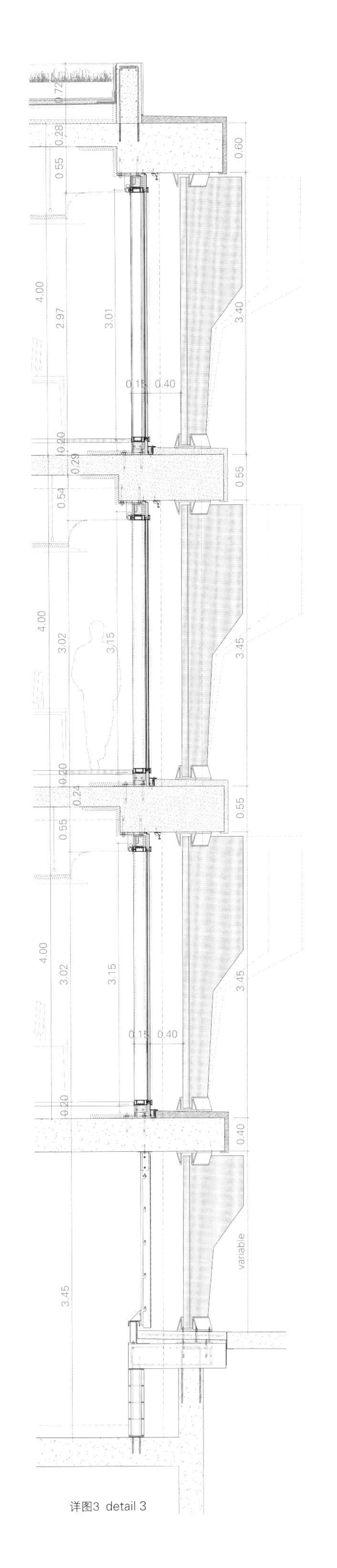
0.72
0.28
0.55
0.60
4.00
2.97
3.01
3.40
0.15
0.40
0.20
0.29
0.55
0.54
4.00
3.02
3.15
3.45
0.20
0.24
0.55
0.55
4.00
3.02
3.15
3.45
0.15
0.40
0.20
0.40
variable
3.45

详图3 detail 3

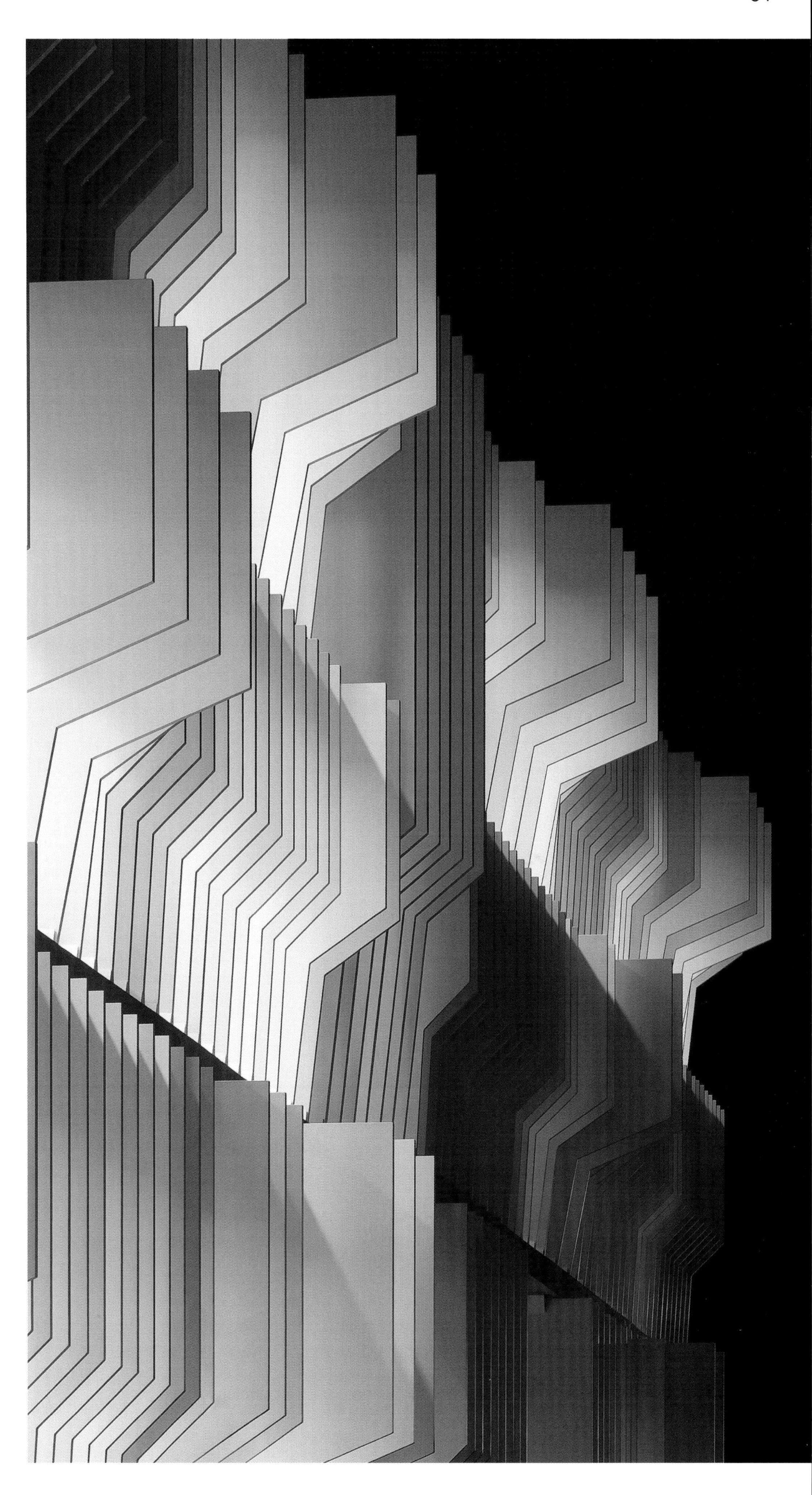

BBVA

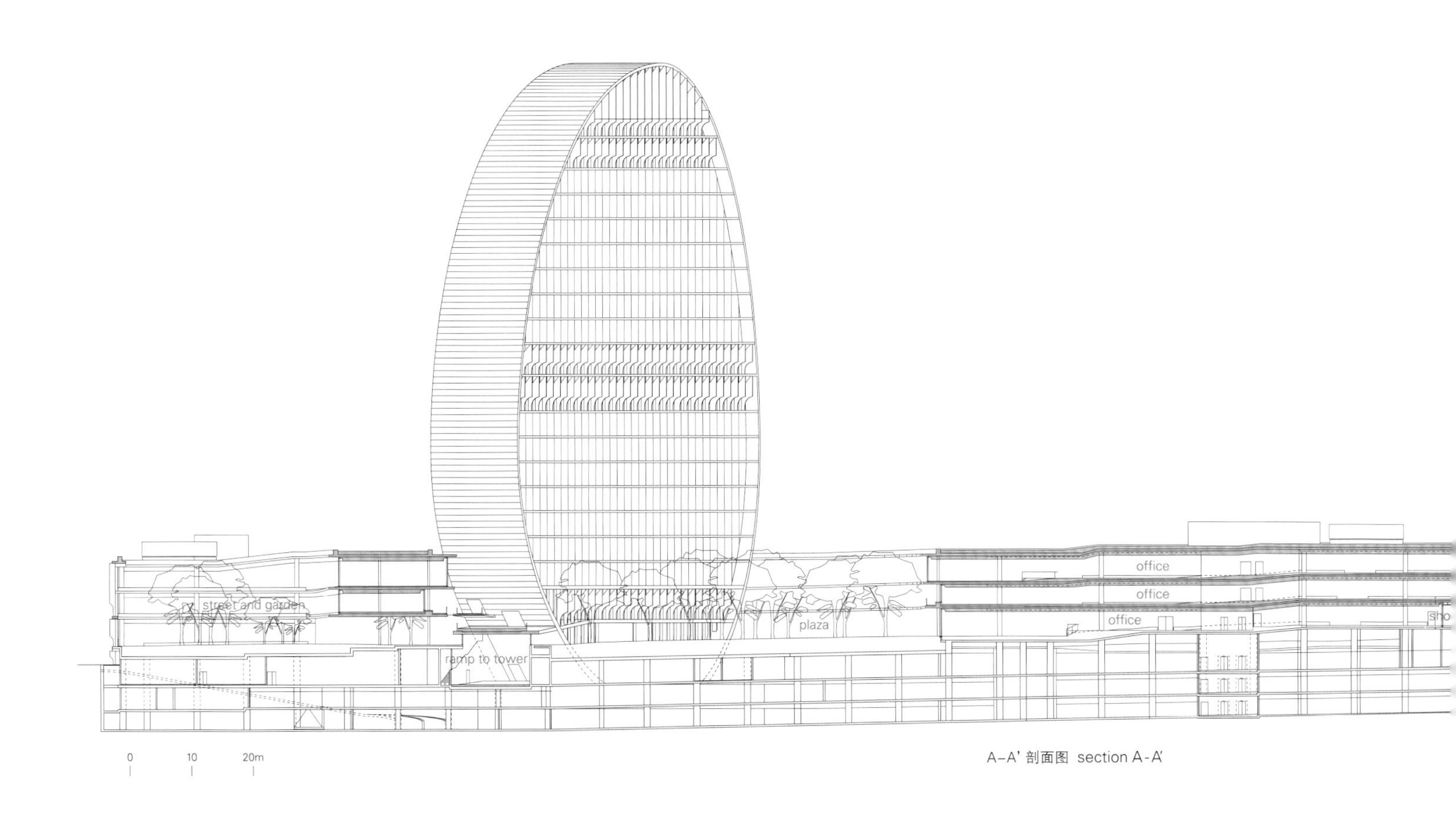

A–A' 剖面图 section A–A'

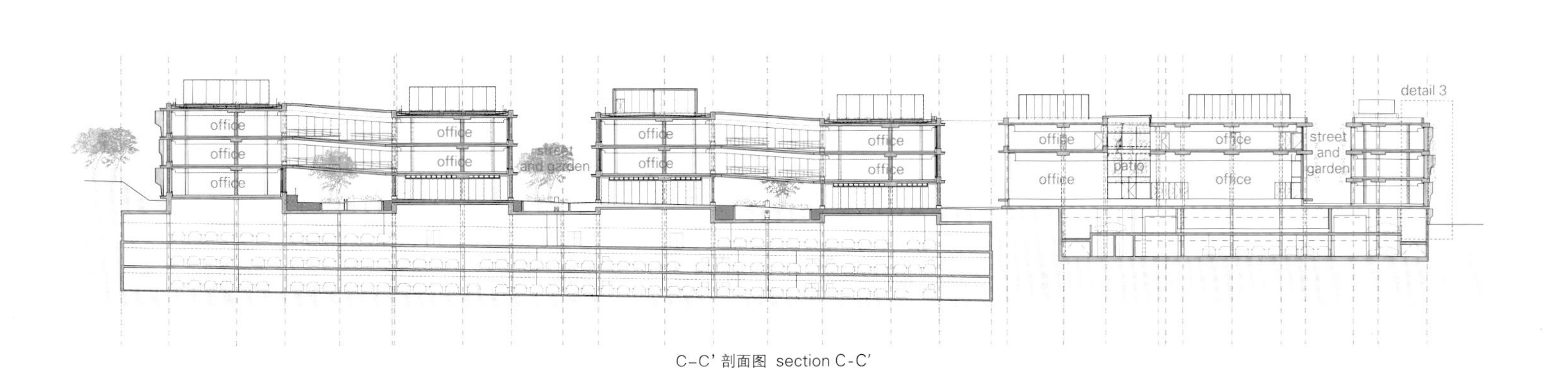

C–C' 剖面图 section C–C'

项目名称：New Headquarters for BBVA / 地点：Calle Azul 4, 28050 Madrid, Spain
建筑师：Herzog & de Meuron / 主管合伙人：Jacques Herzog, Pierre de Meuron; Christine Binswanger, David Koch
项目团队：Nuno Ravara - Associate, Project Director; Miquel Rodríguez - Associate; Stefan Goeddertz - Associate; Benito Blanco, Alexander Franz, Mónica Ors Romagosa, Thomas de Vries, Alexa Nürnberger, Xavier Molina, Enrique Peláez, Nuria Tejerina, Manuel Villanueva, Ainoa Prats / 执行建筑师：Martinez FM Arquitectos, Ortiz y León Arquitectos
总设计：UTE Nueva Sede BBVA, Herzog & de Meuron SL, Drees & Sommer, Martinez FM Arquitectos, Ortiz y Léon Arquitectos
景观设计：Vogt, Benavidez Laperche, Phares, Alvaro Aparicio
机电暖通设计：Arup(London), Arup(Madrid), Grupo JG, Estudio PVI
结构工程：Arup(London), Arup(Madrid), BOMA S.L., INES / 城市设计：Ezquiaga S.L
造价顾问：Drees & Sommer, S.L, Integral S.A. / 艺术总监：Herzog & de Meuron SL
设计项目管理：Drees & Sommer, SL; Integral, S.A / 项目管理：Hill International / 总承包商：Acciona
客户：BBVA - Banco Bilbao Vizcaya Argentaria S.A
用途：office space, auditorium, press room, tradding room, business centers, restaurants, cafeterias
用地面积：59,125m² / 建筑面积：56,635m² / 总建筑面积：251,979m² / 总体积：1,004,775m³ / 立面面积：46,928m²
设计：竞赛设计时间：2007.9—2007.11; 方案时间：2008.9—2009.3; 设计开发时间：2008.9—2009.9
施工：施工文件：2009.1—2010.7; 施工设备：2009.9—2015.12
摄影师：©Rubén P. Bescós (except as noted)

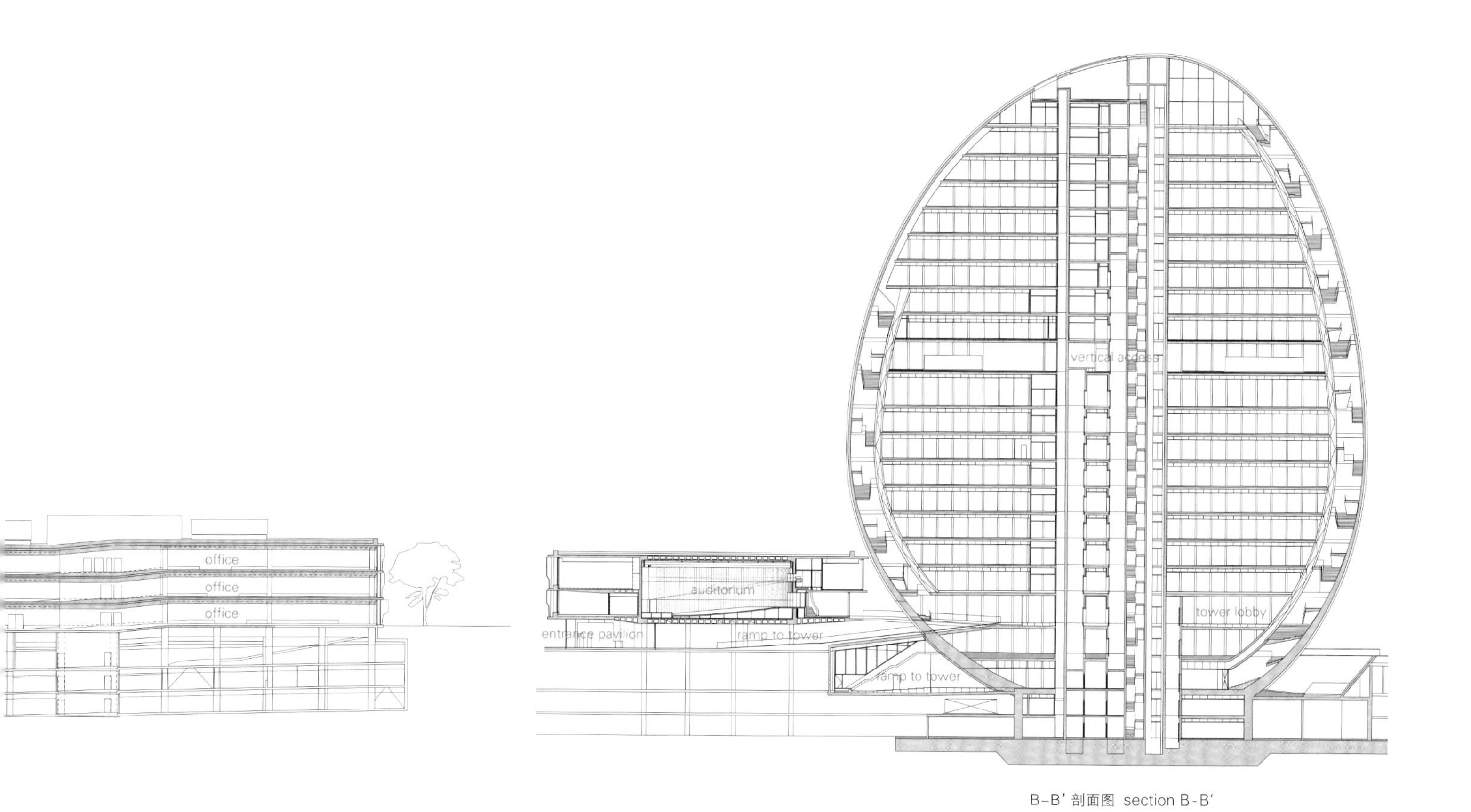

B-B' 剖面图 section B-B'

The new headquarters is designed for 6000 ft. Both the site and the scale of the development challenged us to find a radical solution – we choose to create an inward looking oasis in this otherwise anonymous urban landscape, a place that establishes a balance between the natural and the built and that functions both like a small city and a big garden. The existing buildings are altered to tie in with the new structures, and to create offices and gardens of similar linearity and scale. They are either cut out or filled in to be integrated into the overall "fabric".

A Southern Type of Architecture

It is a raw architecture, one where the structure is prominently expressed. It is a design that is informed by the strong influence of the solar conditions, which ultimately results in a southern type of architecture. Along the rather narrow inner gardens and streets, concrete columns and cantilevering floor slabs provide shade to prevent excessive sun, which reduces demand for air conditioning. The full height but recessed glazing provides good daylight conditions in the offices in order to minimize artificial lighting. Along the periphery of the complex we developed brise-soleils that are fixed in between the floor slabs. Unlike the prominent modern references, these are cut out in the lower part at an angle to provide more view and daylight where protection is needed least- resulting in figurative element that vary in direction and size according to the solar angle and program. The sloping site creates another subtle yet influential consequence on the facade as the brise-soleils adjust in height.

Locating BBVA in the Madrid Skyline

A round like plaza is cut out of the carpet, and then, it is as if this mass were tilted upward to become a very slim tower to mark BBVA in the Madrid skyline. In contrast to the low-rise offices, the tower offers another type of workspace, with views across the city and to the mountains. The plaza is planted with hundreds of trees and surrounded by various communal facilities. Together, the plaza and the tower provide orientation to the entire complex.

韩泰轮胎中央研发中心
Hankook Technodome

Foster + Partners

创新之路上的创造

位于韩国大田的韩泰轮胎研发中心——亚洲硅谷的心脏——在过去20多年里一直是该公司的技术中心。该公司鼓舞人心的技术催生了防爆胎轮胎和密封保护轮胎，为韩泰轮胎转型为韩国顶级轮胎制造商和全球顶级供应商奠定了基础。

公司的发展远远超出了预期，因此，研发实验室的容量很快达到了顶峰。这样一来，公司就需要改善研究基础设施和工作环境。研发中心的目标是扩大研发能力，打造一个智能的研发工作环境。它既要解决实际的研究需求，又要发挥全球研发网络中心的作用，最重要的是，它要成为韩泰轮胎创新的象征。

体现抽象设计的高科技橱窗

未来主义高科技建筑大师诺曼·福斯特是首选设计师。他的位于伦敦的设计公司Foster ＋ Partners擅长挑战建筑的极限，刚好与韩泰轮胎公司的发展目标完美吻合：成为一家追求未来主义形象的前沿科技公司。中央研发中心是福斯特在韩国做的第一个项目。韩泰轮胎公司希望该中心通过使用未来主义和科学的白色空间，与公共场合的黑色橡胶块形成鲜明的对比，能够精确地代表该公司的形象。

Technodome（中央研发中心），顾名思义，是一种圆顶状的建筑。有许多曲线不仅存在于屋顶上，而且在建筑内部也有。韩泰轮胎公司拒绝简单直白的诠释方法，即曲线是由轮胎的圆形形状演绎出来的。虽然避免了直接引用公司的标志，比如，建筑上的标志，但他们也希望公司的历史、工作原理和企业文化能象征性地体现在建筑设计上。

该项目的愿景是创建一个光滑、现代而又神秘的建筑，带有漂浮的银色屋顶。屋顶应同时提供遮阳功能，并允许适量的光线进入室内。在整个仿真和微调的过程中，建筑师对屋檐的曲线进行了精确的反复调整。福斯特通过利用太阳纬度的差异，找到了提供自然光、供暖和制冷效果的完美形状。

硬件引领软件创新

新研发中心的核心是轮胎测试和研究实验室，向受邀的参观者和工作人员展示，以强化韩泰轮胎公司的核心身份。该项目占地9.8万平方米，旨在吸引业内顶尖人才，为他们提供一个能够激发灵感的工作场所，以及光线充足的办公室、先进的实验室和充满活力的社交空间，以培养开放和创新的文化。它鼓励人们在工作场所内进行非正式的互动，这里还有用于自发组建团队会议的中心会议室。

对现有建筑的分析提供了对实验空间的深入了解——这些实验室中有三分之一需要隔离维修区，因此必须位于一层，而其余的空间需要位于两层高的区域，以容纳设备。建筑的剖面设计是解决这个复杂空间难题的关键。建筑呈五指状排列，办公与工业单元平行。除了创建一个综合型的动态平面来促进不同区域之间的视觉联系之外，这种安排还非常灵活，可以支持未来的使用变化。每根“手指”之间的间隙将日光引入楼面板的中心。从侧面看去，由于邻近政府场地存在高度限制，所以为了表示呼应，楼层从四层上升到七层。

研究空间沿着顶部照明的中央脊柱延伸，从南面的餐厅和入口一直延伸到北面的员工宿舍。椭圆形的玻璃会议室悬挂在通高空间的上方，吸引阳光穿过建筑。交通流线的设计策略在公共区域和较为敏感的产品开发区之间创造了自然的分割，让游客们可以从作为最新系列产

OUT

IN
P

品展示空间的大厅开始参观建筑，沿着可以看到试验区的中央脊柱前行，最后再走到外面的公园。所有这些不同的功能都统一布置在一个有着大胆设计姿态的、宽大的银色屋顶天篷之下。

Created on the Road to Innovation

Hankook Tire's Research and Development Center in Daejeon, South Korea – the heart of Asia's Silicon Valley – has been company's technology hub for over two decades. The company's inspiring technology gave rise to run-flat tires and seal-guard tires, which laid the foundation for Hankook Tire's transformation into the nation's top tire manufacturer and also a top-tier supplier on the global stage.

The company grew far beyond expectations, and as a result, the R&D lab quickly reached its peak capacity. This led to the need for improving the research infrastructure and working environment. The R&D center's goal was to expand research capabilities and create a smart R&D working environment. It would have to handle actual research demands and at the same time play the role of global R&D network hub, and above all, serve as a symbol of Hankook Tire's innovation.

High-tech Showcase Embodying Abstract Design

Norman Foster, a master of futuristic high-tech architecture was the very first choice. His London-based firm, Foster + Partners' challenging the limits of architecture was a perfect fit for Hankook Tire's pursuit of a futuristic image as a cutting-edge technology company. The Technodome is the first Normal Foster's project built on Korean. Hankook Tire wanted the center to accurately represent the company via the use of futuristic and scientific white spaces that contrasted with the black rubber lumps branded in the public. The Technodome is, as its name indicates, a building in the form of a dome. There are many curves not only on the roof but also inside the building. Hankook Tire refused the one-dimensional interpretation that the curves are derived from the circular shape of tires. While avoiding symbols directly

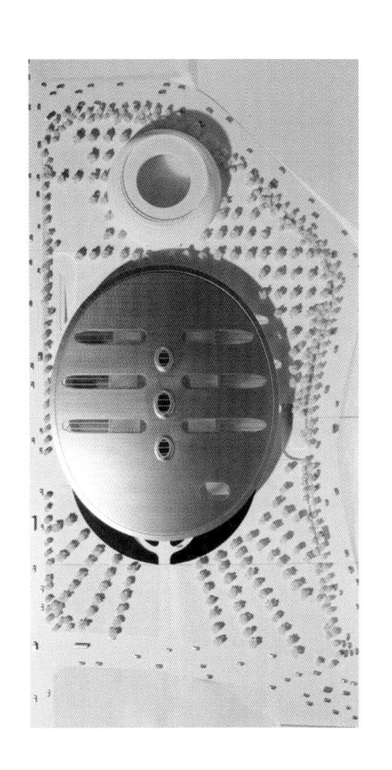

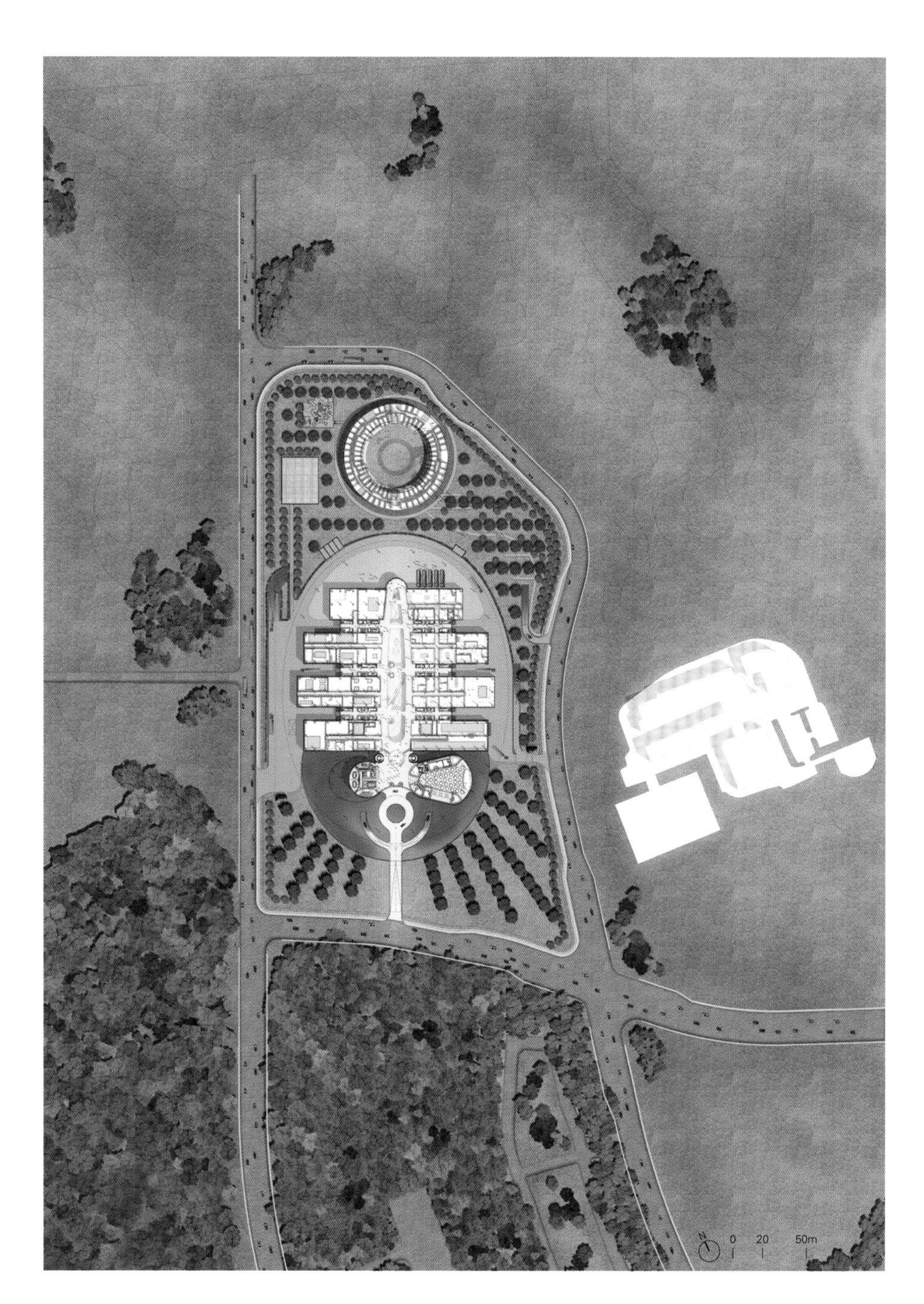
0
20
50m

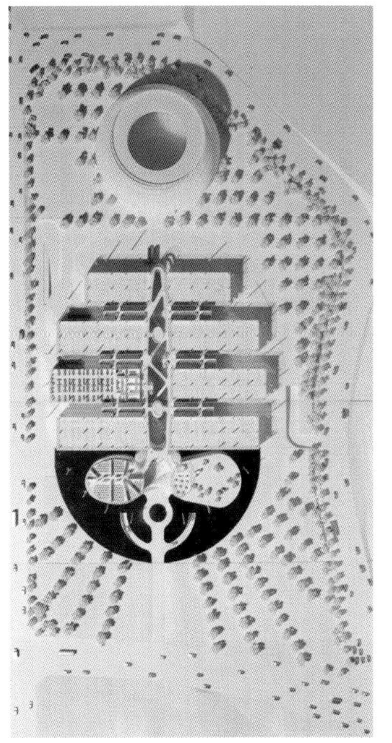

项目名称：Hankook Tire, Central R+D Center / 地点：Daejeon, Korea
建筑师：Foster + Partners
设计团队：Norman Foster, David Nelson, Spencer de Grey , Nigel Dancey Giles Robinson, Iwan Jones, Stefano Cesario, Miguel Costa, Sanhita Chaturvedi, Joost Heremans, Chris Johnstone, Sharat Kaicker, Kristi Krueger Angel Sanchez Lazcano, Huasop Lee Sarah Lister, Eva Palacios, Isidora Radenkovic Paola Nena Sakits Nicola Scaranaro, Milena Stojkovic, Vincent Westbrook
工程团队：Roger Ridsdill Smith, Piers Heath, Emma Clifford, Anis Abou Zaki, Ricardo Candel Gurrea, Dimitra Kyrkou, Carole Frising, Rob Slater
当地合作建筑师：SAMOO / 结构工程师：CS Structural Engineering
土木工程师：Saegil / 电气工程师：SEC / 设备顾问：HIMEC Corporation
施工经理：ITM Corporation / 照明工程师：Alto / 室内设计：Dawon
景观设计顾问：Solto Associates / 消防顾问：Korea Fire Protection UBIS
可持续性顾问：Samoo / 客户：Hankook Tire Co. Ltd.
总用地面积：70,387m² / 总建筑面积：96,328m² / 施工时间：2014 / 竣工时间：2016
摄影师：
©Nigel Young (courtesy of the architect) - p.78~79, p.84~85 left-top, left-bottom, p.86~87 left-top, left-bottom
©Kim Jong-oh - p.76, p.82, p.83, p.84~85 right, p.86~87 right, p.88~89
Courtesy of the Hankook Tire Co. Ltd. - p.72~73, p.74~75

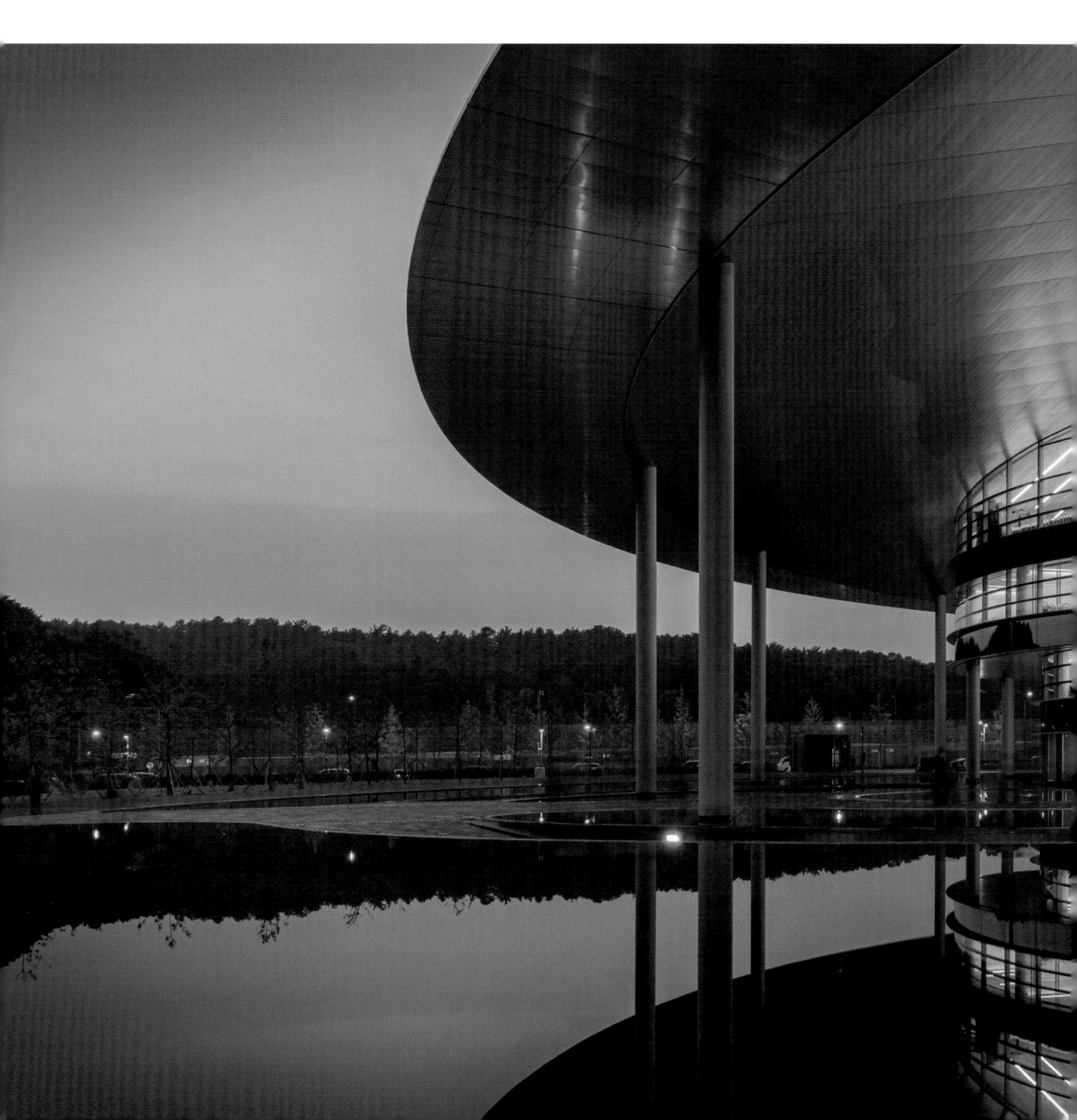

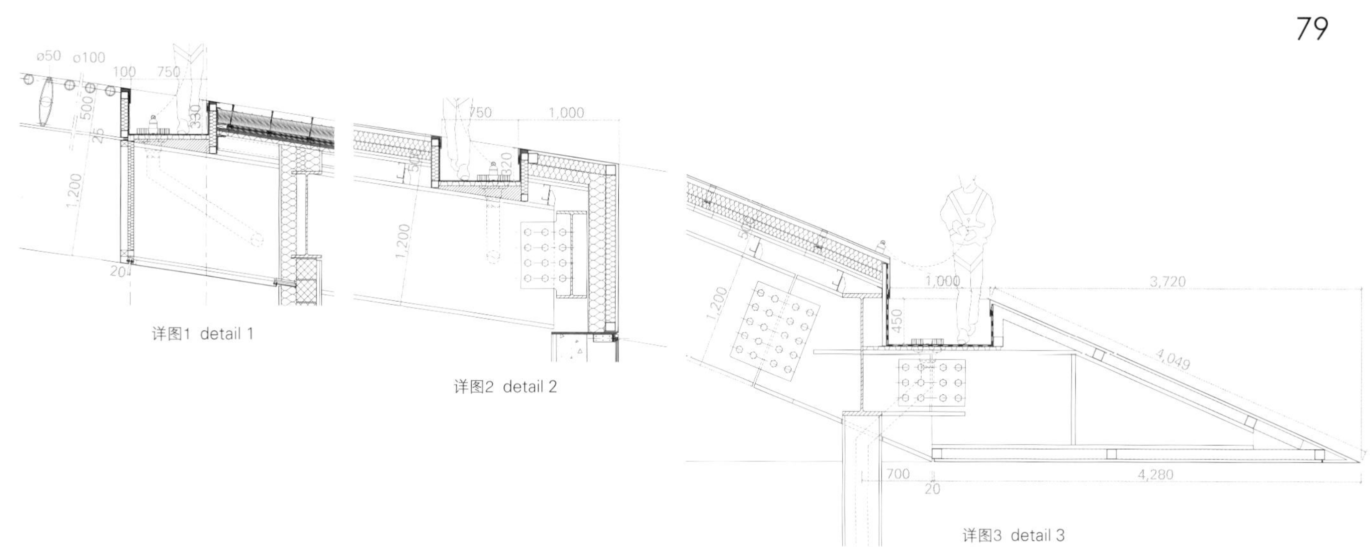

详图1 detail 1

详图2 detail 2

详图3 detail 3

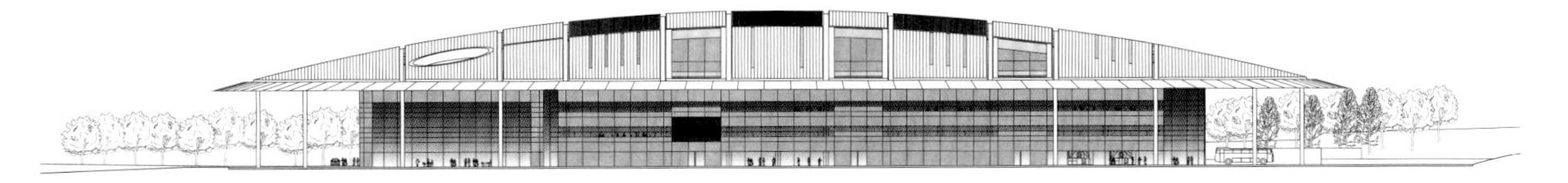

西立面 west elevation

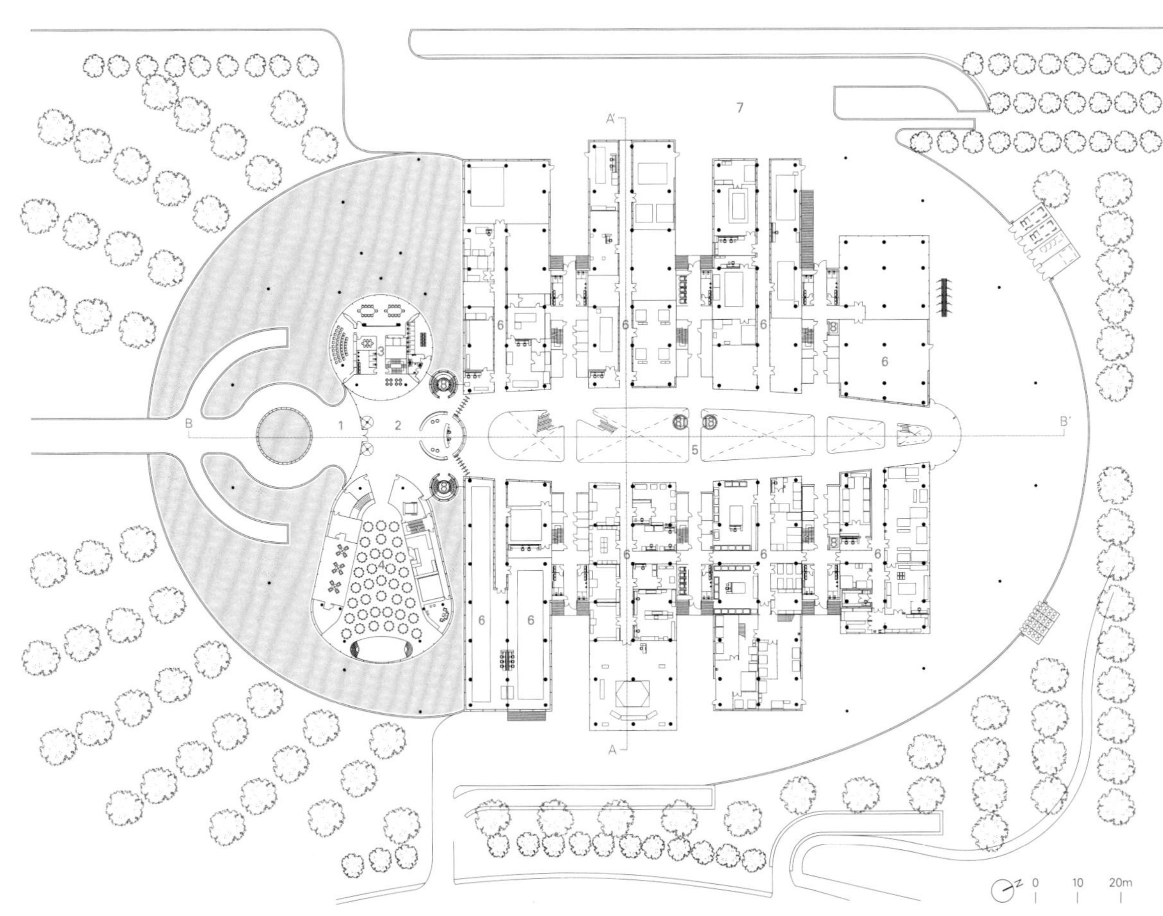

一层 ground floor

1. 下客处
2. 大厅
3. 商务中心
4. 礼堂
5. 中央空间
6. 实验室
7. 设备庭院
8. 电梯

1. drop off
2. lobby
3. business center
4. auditorium
5. central space
6. laboratories
7. service yard
8. lift

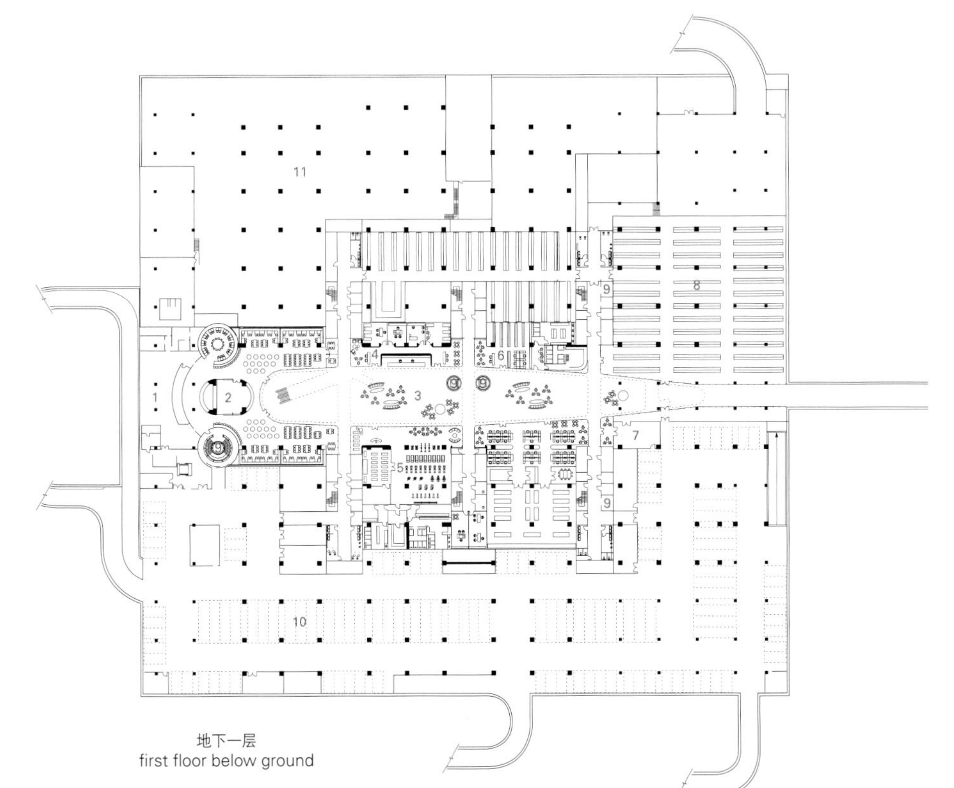

地下一层
first floor below ground

1. 厨房
2. 员工餐厅
3. 中央空间
4. 保健室
5. 健身房
6. 图书馆
7. 收发室
8. 轮胎库房
9. 电梯
10. 停车场
11. 设备间

1. kitchen
2. staff cafeteria
3. central space
4. healthcare
5. fitness
6. library
7. postroom
8. tire storage
9. lift
10. car park
11. plant

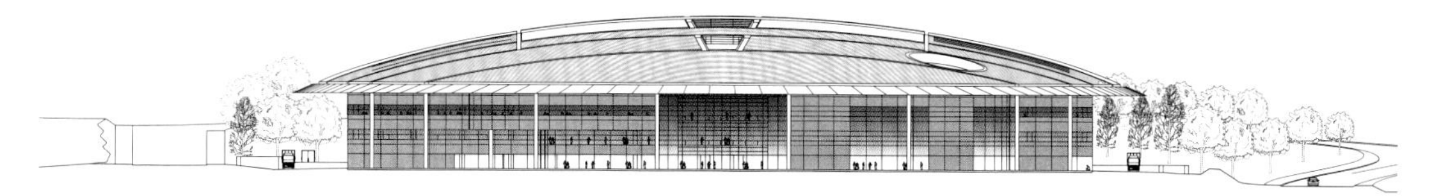
南立面 south elevation

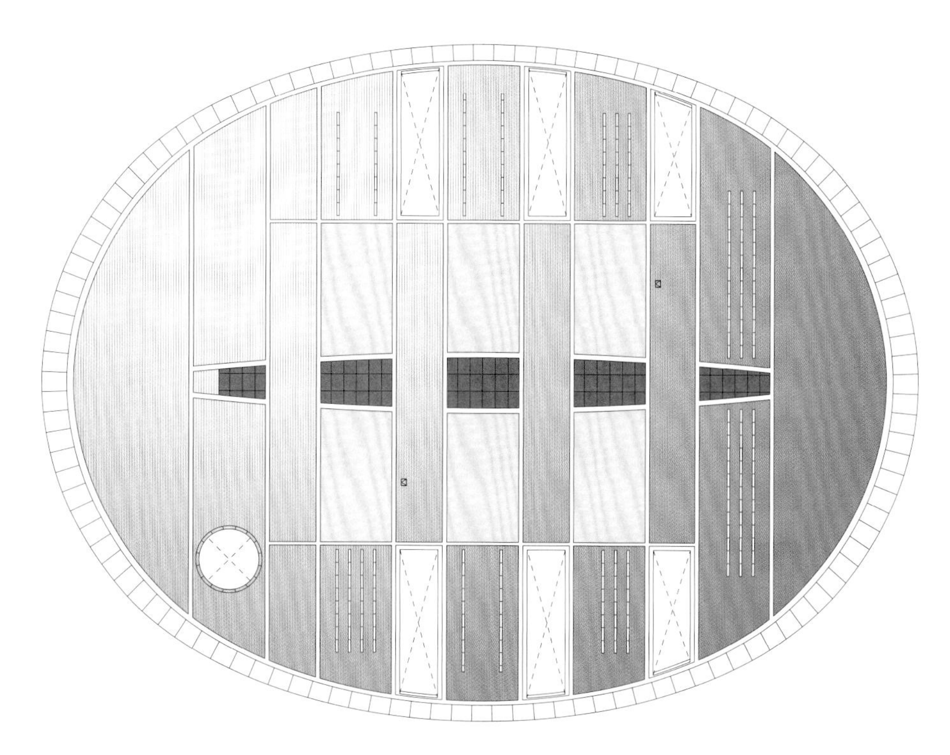
屋顶 roof

1. 教室
2. 咖啡厅
3. 花园
4. 会议室
5. 中央空间
6. 办公室
7. 行政区
8. 实验室
9. 电梯

1. academy
2. cafe
3. garden
4. meeting pod
5. central space
6. offices
7. executive zone
8. laboratories
9. lift

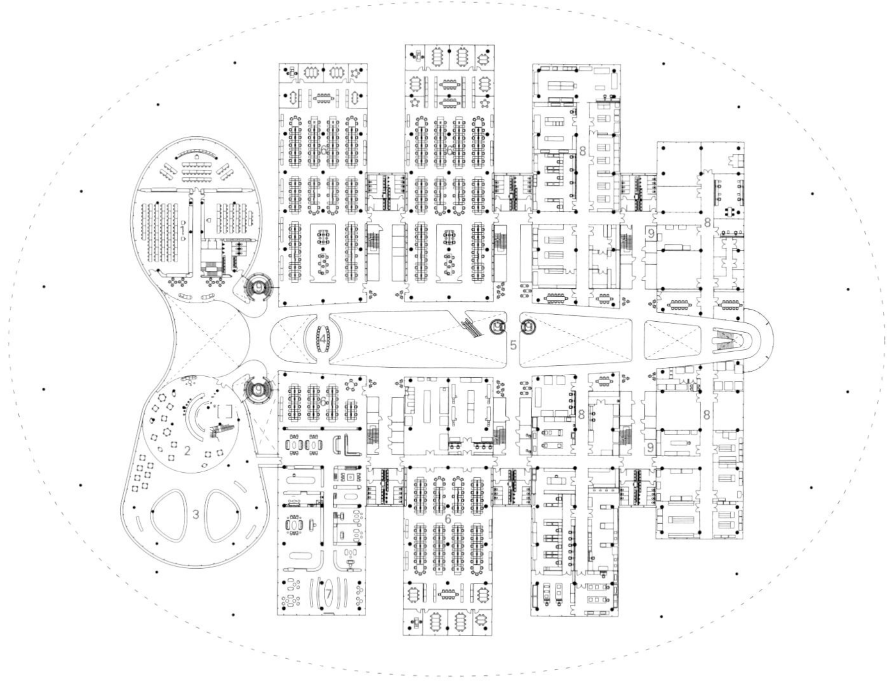
五层 fourth floor

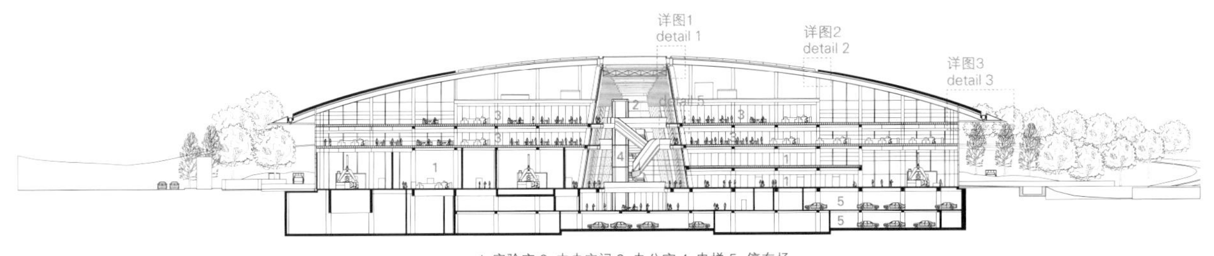

1. 实验室 2. 中央空间 3. 办公室 4. 电梯 5. 停车场
1. laboratories 2. central space 3. office 4. lift 5. car park
A–A' 剖面图 section A-A'

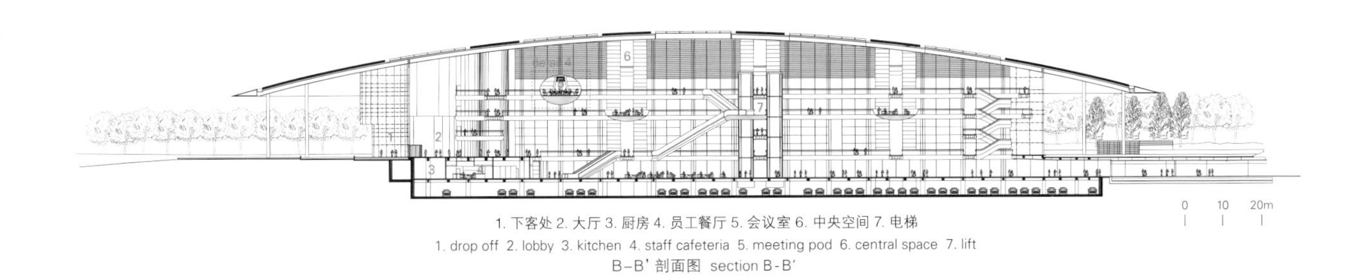

1. 下客处 2. 大厅 3. 厨房 4. 员工餐厅 5. 会议室 6. 中央空间 7. 电梯
1. drop off 2. lobby 3. kitchen 4. staff cafeteria 5. meeting pod 6. central space 7. lift
B–B' 剖面图 section B-B'

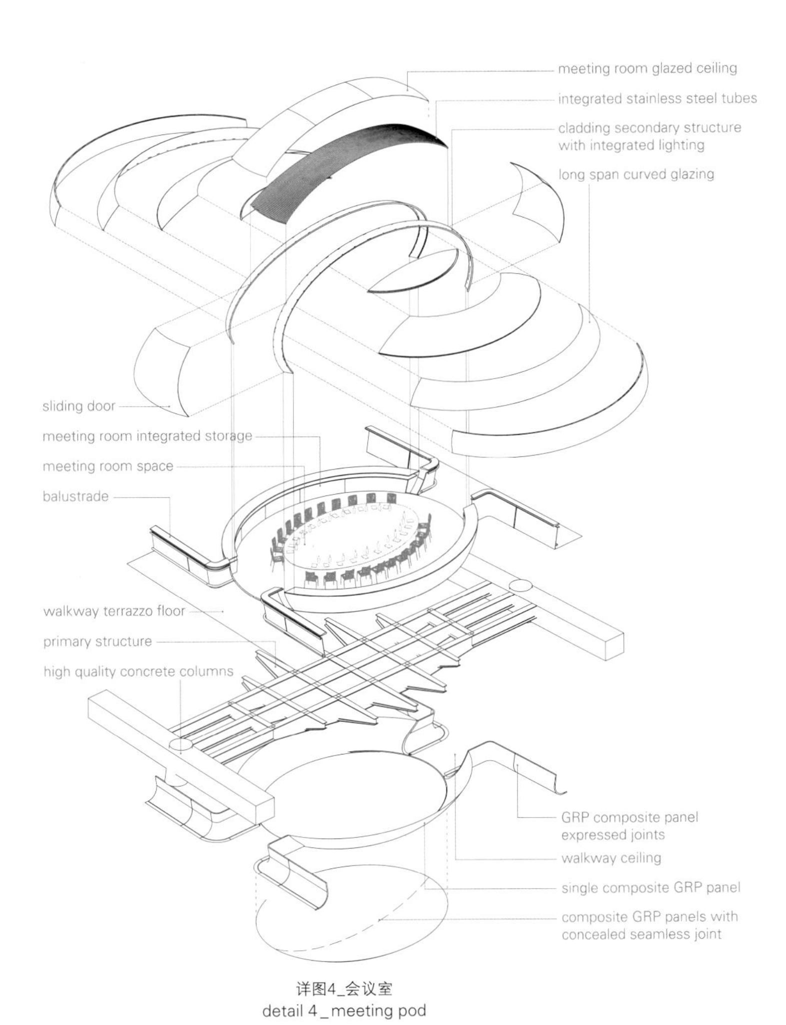

详图4_会议室
detail 4_meeting pod

Pillar
3

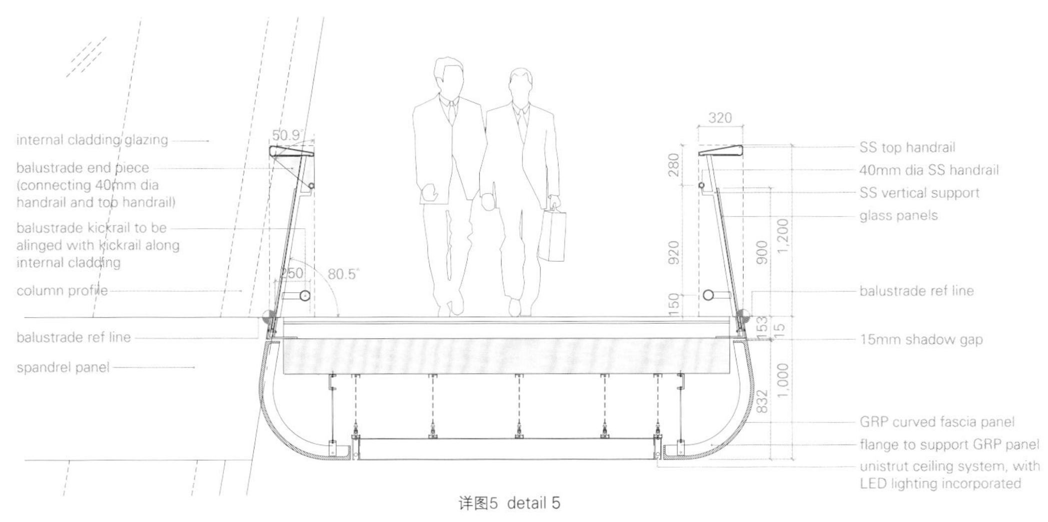

internal cladding glazing
balustrade end piece
(connecting 40mm dia
handrail and top handrail)
balustrade kickrail to be
alinged with kickrail along
internal cladding
column profile
balustrade ref line
spandrel panel
50.9
250
80.5°
320
280
920
150
900
1,200
153
15
1,000
832
SS top handrail
40mm dia SS handrail
SS vertical support
glass panels
balustrade ref line
15mm shadow gap
GRP curved fascia panel
flange to support GRP panel
unistrut ceiling system, with
LED lighting incorporated

详图5 detail 5

Pillar

Bistrofine

referencing the company, such as a logo adorning the building, they wanted the company's history, working principles and corporate culture symbolized architecturally.

The vision was to create a sleek, contemporary and mysterious building with a floating silver roof. The roof should simultaneously provide shade and allow an appropriate amount of light into the interior. The curving lines of the eaves were precisely adjusted repeatedly throughout the simulation and fine-tuning processes. Foster found the perfect shape to provide natural light, heating and cooling by applying the difference in latitude of the sun.

Hardware Leading Innovation of Software

The centrepiece of the new R&D facility are the tyre testing and research laboratories, on display to invited visitors and staff to reinforce Hankook's core identity. The 98,000m² facility aims to attract the industry's top talent, providing an inspirational place to work, with light filled offices, advanced laboratories and dynamic social spaces to nurture a culture of openness and innovation. It lays emphasis on informal interaction within the workplace, with central meeting pods for spontaneous team meetings.

Analysis of the existing buildings provided an insight into the testing spaces – one third of these laboratories require isolation pits, so have to be located on the ground floor, and the rest need to be located in double-height areas to accommodate equipment. The building's section was key to resolving this complex spatial puzzle. The building is arranged as five fingers, with parallel office and industrial units. As well as creating a dynamic, integrated plan that promotes visual connections between different areas, the arrangement is highly flexible to enable future changes in use. Breaks between each finger draw daylight into the heart of the floor plate. In profile, the levels step up from four to seven storeys, in response to the height restriction imposed by an adjacent government site.

The research spaces extend along a top-lit central spine that runs from the restaurant and entrance in the south to the staff accommodation to the north. Glazed oval meeting pods are suspended above the full-height space, which draws daylight through the building. The circulation strategy creates a natural divide between public areas and more sensitive product development zones, allowing visitors to follow a tour through the building, from the lobby, which functions as an exhibition space for the latest product range, along the spine with views into the testing areas, to the parkland outside. All these different functions are unified beneath the bold gesture of a wide silver roof canopy.

可隆公司及研发总部

Kolon Corporate and Research Headquarters

Morphosis

总部位于韩国首尔的可隆集团是一家多元化的公司，其业务范围从纺织品、化学品和可持续技术，到运动服装和成衣时尚市场的原创服装系列，涵盖甚广。可隆集团的38个部门涵盖了研究、主要材料制造和产品建设，其独特的部门配置使公司能够充分利用自己的资源和优势获得发展，并在部门之间建立创新的合作关系。支持这种合作模式是可隆新公司总部及研发大楼设计背后的主要目标。该建筑将研究人员、领导和设计师聚集在同一个地点，将灵活的实验室设施与行政办公室和活跃的社交空间相结合，鼓励在整个公司范围内进行更大的互动和交流。

公司总部位于麻谷区，这是一个新兴的技术和轻工业中心，正在振兴首尔东南部的汉江地区。在首尔市政府的推动下，麻谷区被设想为一个"工业生态系统"，在这里一系列的技术和信息领域将共同定位，创造新的交叉市场。可隆是第一批将公司总部和研发业务搬到麻谷的公司之一，其新总部大楼将为该地区的建筑性能和设计设定标准。该项目占地四英亩（约合1.62ha），毗邻麻谷的中央公园——这里将坐落该地区第一座主要的完工建筑。

这座建筑朝向公园折叠，为较低的楼层提供了被动遮阳。该折叠式体量连接了三座延伸的实验室翼楼，包含会议室和社交空间，并在街道上增加了旗舰零售店和展览馆，向公众传达品牌的愿景。透明的地面将景观延伸到室内，将光线和运动引向开放的行人通道和宏伟的入口。高30m、长100m的宽阔多层中庭作为建筑的社交中心。中庭的透明内衬系统由一个巨大的8m长的延伸结构组成，它可以不断变化展示可隆自己生产的面料，从中庭也可以看到所有楼层的动态。该公司的管理策略强调毫无阻碍的沟通，譬如，通过30m高、100m长的主楼梯可以看到建筑内部。主楼梯连接着二层到四层，它不仅是一条直接的通道，有时也作为一个活动舞台，还可以作为员工放松的地方。换句话说，开放的中庭展示了所有楼层的运动，使空间成为建筑的社交中心。楼梯周围的墙壁上密密麻麻地装满了300多个菱形模块，每一个模块都展示着各种各样的产品，例如，由可隆众多子公司生产的安全气囊、汽车安全座椅、步行服、人造皮革和人造草坪。这使得墙壁本身成为一个巨大的展廊。

这座建筑的性能被视为一个整体概念，包括能源效率、资源节约和环境管理，与教育、员工健康和福祉相互协调。除了达到韩国最严格的可持续发展认证目标之外，该项目还通过屋顶露台、庭院及其他措施为员工增加了获得自然光线和空气的机会，重点关注工作环境的质量。其他可持续措施包括：绿色屋顶；回收材料；使用泡沫甲板，可以减少30%的混凝土用量。此外，严格的能效模拟使西侧的公共空间成为一个净零能耗区，由建筑内部产生的能源维持。西侧立面独特的百叶窗系统既体现了建筑的性能，也成为一种象征。立面由超过400个单元组成，这种参数化的形状使人们能将公园的景色尽收眼底，同时阻挡了来自西部的阳光直射。这些巧妙而又独特的单元错综复杂地连接在一起，设计得像紧密编织的织物，传达出纺织的主题，这代表了公司在纺织行业的根基，并通过在建筑立面本身的展示将图像带到最前端。这些单元由玻璃纤维增强塑料制成，使用了可隆公司自己的高科技面料之一——芳纶，以大幅提高材料的抗拉强度。总而言之，建筑的选址、空间品质和技术创新均体现了可隆公司在可持续发展方面的投资和承诺。

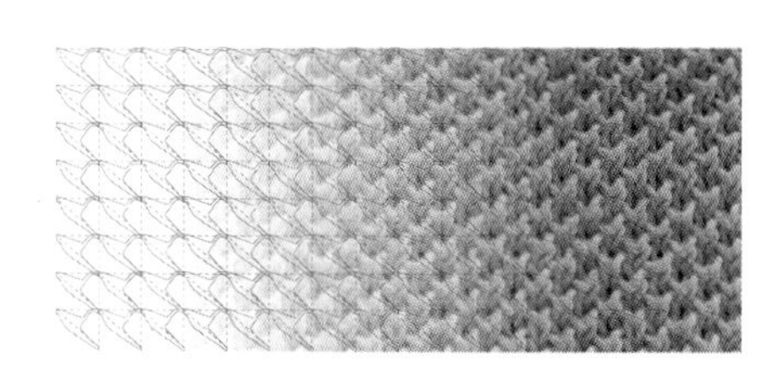

立面概念
facade concept

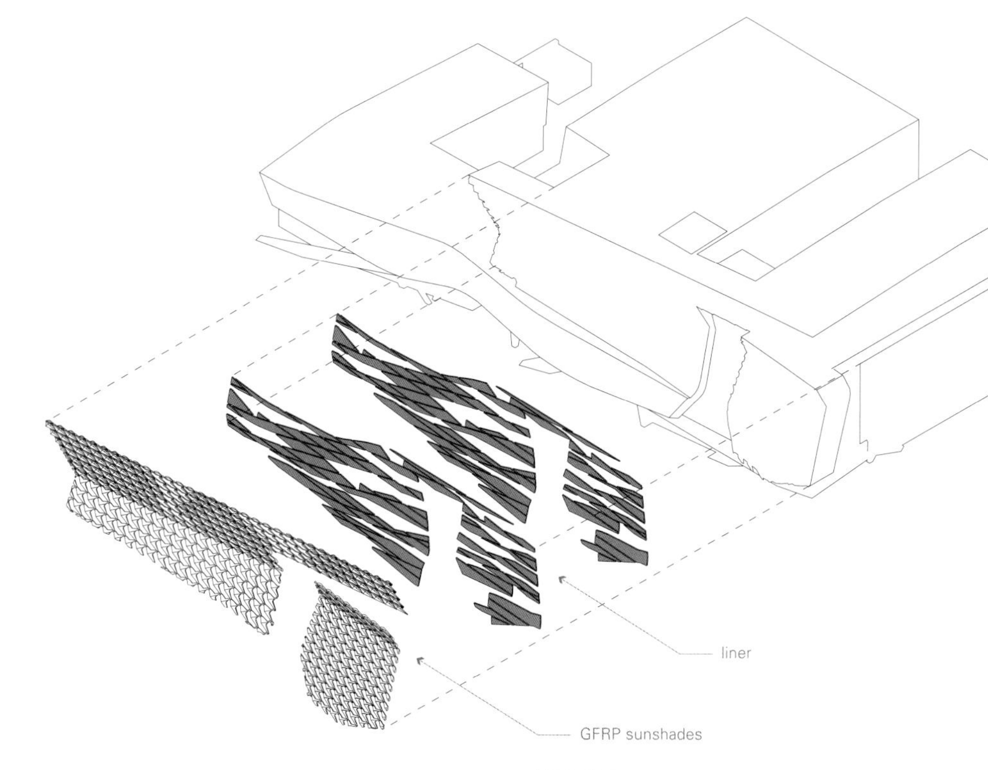

西立面分层
west facade layers

The Kolon Group, based in Seoul, is a diverse corporation whose activities range from textiles, chemicals, and sustainable technologies, to original clothing lines in the athletic and ready-to-wear fashion markets. Between the group's 38 divisions, Kolon covers research, primary material manufacture, and product construction – a unique configuration that enables the company to capitalize on its own resources and advances, and to forge innovative collaborations between divisions. Supporting this collaborative model was a primary goal behind the design of Kolon's new corporate headquarters and research facilities. Bringing researchers, leadership, and designers together in one location, the building combines flexible laboratory facilities with executive offices and active social spaces that encourage greater interaction and exchange across the company.

The headquarters is located in the Magok district, an emerging hub for technology and light industry that is revitalizing the Han-River area in south-eastern Seoul. Fostered by the Seoul Metropolitan Government, the Magok district is conceived to function as an "industrial ecosystem" where a range of tech and information fields will co-locate to spawn new intersecting markets. Kolon is one of the first firms moving their corporate headquarters and R & D operations to Magok, and the new building will set the standard for performance and design in the district. The four-acre project site sits adjacent to Magok's central park – a prominent location for what will be the district's first major completed building.

The building folds towards the park, providing passive shading to the lower floors. Bridging the three extending laboratory wings, this folding volume contains conference rooms and social spaces, augmented by flagship retail and exhibition galleries at the street level to communicate the brand's vision to the public. A transparent ground plane extends the landscape into the interior, drawing light and movement towards an open pedestrian laneway and grand entry. At 30m tall and 100m long, the expansive multi-story atrium serves as the building's social center. The atrium's transparent liner system, which is comprised of massive, 8m 'stretchers' that allow for a changing display of Kolon's own fabrics, grants a full view of movement on all floors. The company's management strategy, which emphasizes

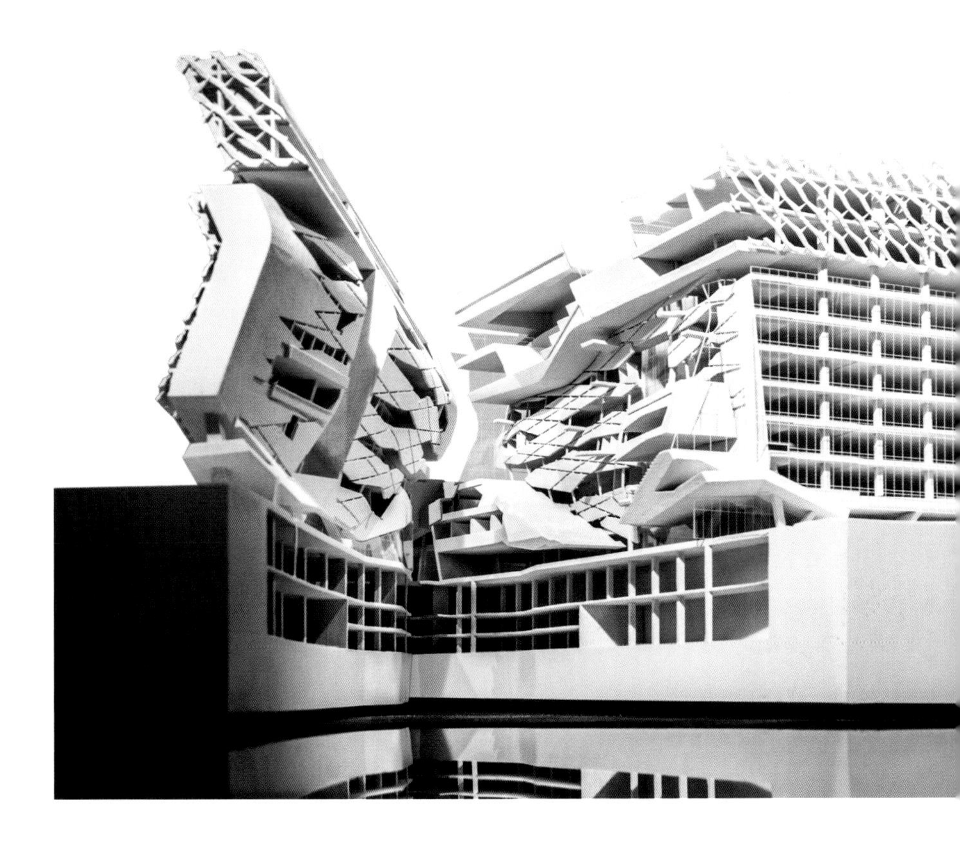

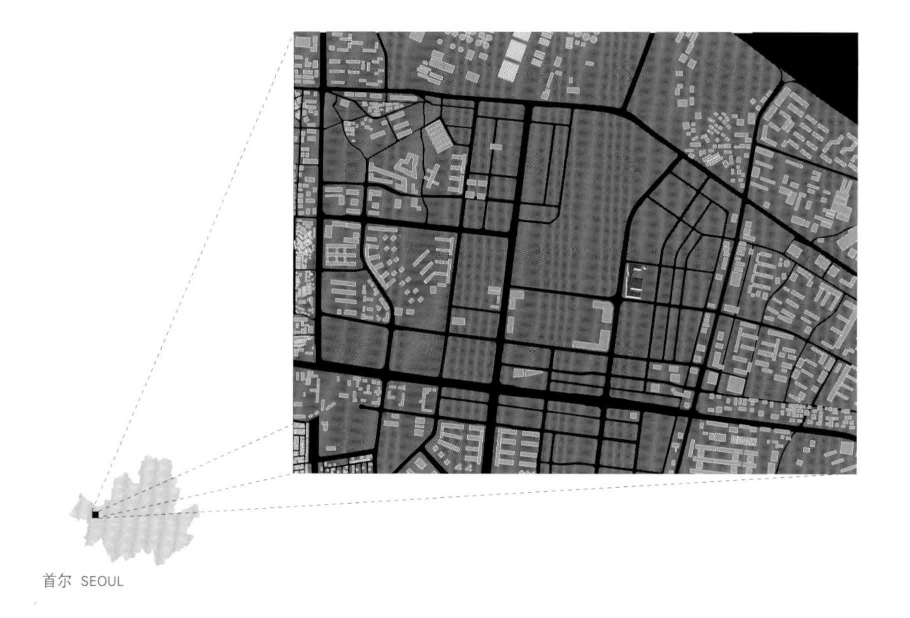
首尔 SEOUL

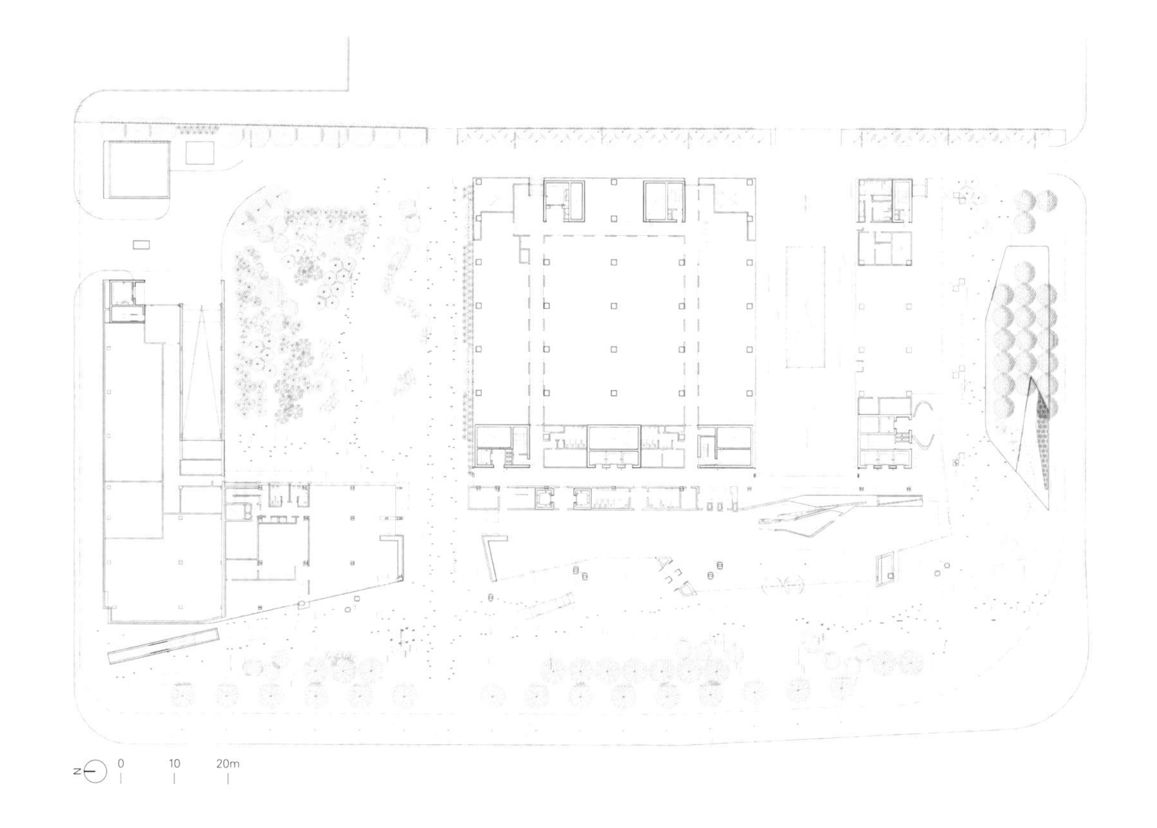

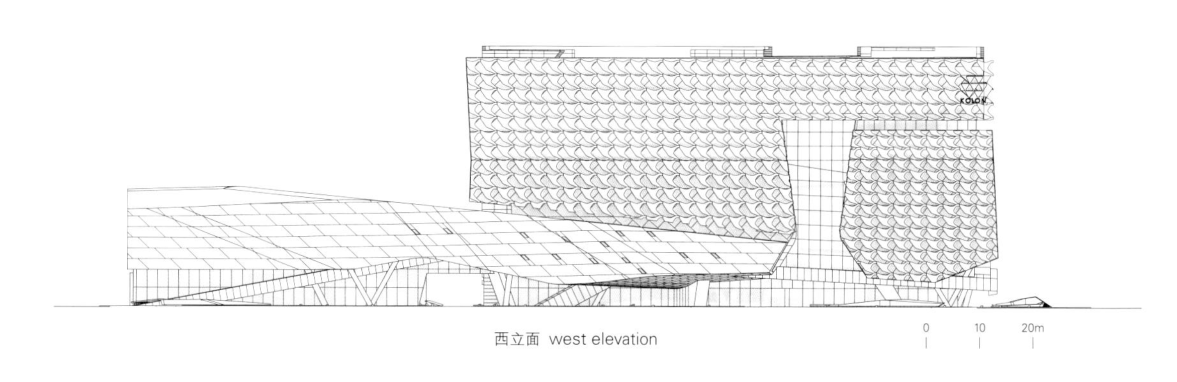

西立面 west elevation

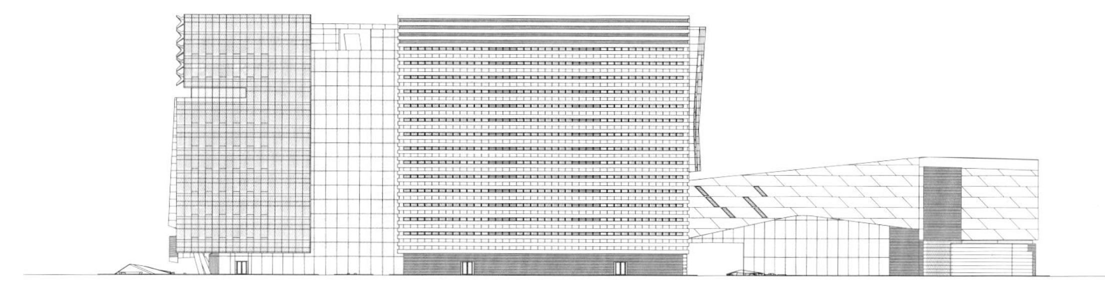

东立面 east elevation

北立面 north elevation

南立面 south elevation

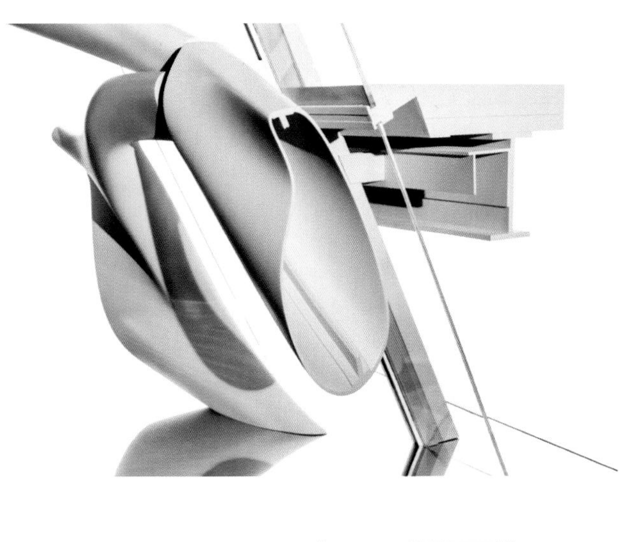

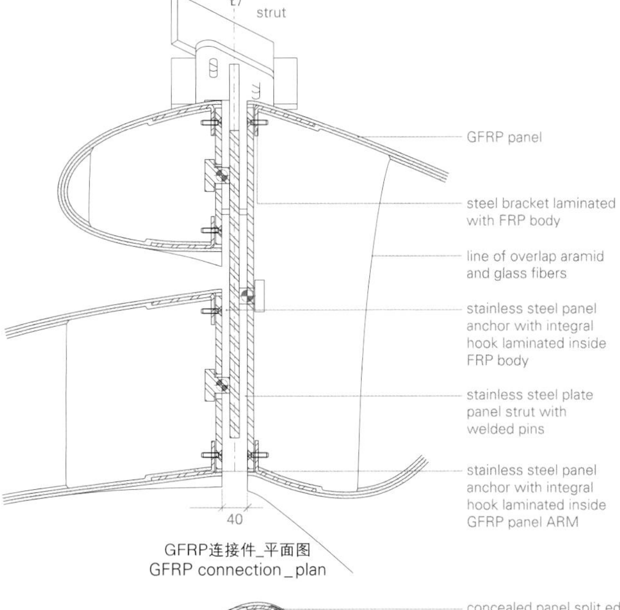

GFRP连接件_平面图
GFRP connection _ plan

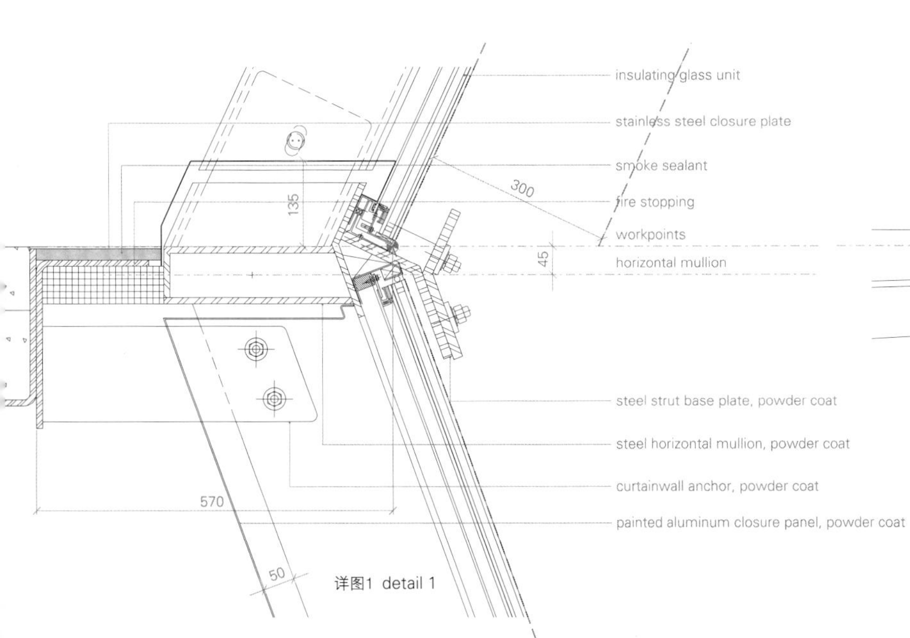

详图1 detail 1

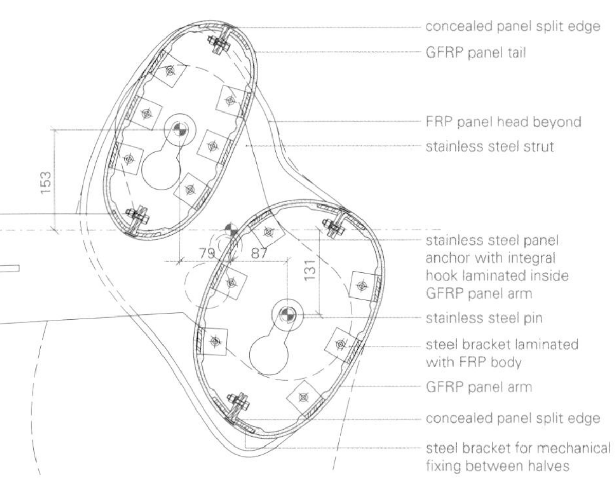

GFRP连接件_尾部和臂部
GFRP connection _ tail and arm

unhindered communication, is represented by the building interior, visible through the 30m tall, 100m long main staircase. The main staircase, which connects floors 2 through 4, acts not only as a direct passage but also at times as an event stage, and at times as a place of relaxation for employees. In other words, the open atrium reveals movement on all floors, making the space the social center of the building. The walls surrounding the staircase is densely packed with over 300 rhombus-shaped modules, with each one displaying a variety of products like air bags, car seats, hiking wear, artificial leather, and artificial turf produced by Kolon's many subsidiaries. This makes walls themselves a massive gallery.

The performance of the building was approached as a holistic concept encompassing energy efficiency, resource conservation, and environmental stewardship, working in concert with education and employee health and well-being. Along with goals the most rigorous sustainability certification in Korea, the project focuses on the quality of the work environment through roof terraces, courtyards, and other measures that increase access to natural light and air for employees. Other sustainable measures include: green roofs; recycled materials; and utilizing a bubble deck slab that reduces the amount of concrete used by 30%. In addition, rigorous energy efficiency simulations allowed the communal space on the west side to be constructed as a net zero energy consumption zone, sustained by energy internally generated by the building. The distinctive brise-soleil system on the western facade is both a performative and symbolic feature of the building. The facade is composed of over 400 units parametrically shaped to allow a view of the park while blocking off direct sunlight from the west. The subtly unique individual units are intricately connected, designed to resemble tightly-woven fabric to convey the textile motif, which represents the company's roots in the textile industry and brings the imagery to the forefront by displaying it on the building facade itself. The units are made from a GFRP formulation that uses one of Kolon's own high-tech fabrics, Aramid, to dramatically increase the material's tensile strength. Together, the building's siting, spatial qualities, and technological innovations express Kolon's investment in and commitment to sustainability.

项目名称：Kolon Corporate and Research Headquarters / 地点：Magok Industrial Park, Seoul, Korea / 建筑师：Morphosis / 设计主管：Thom Mayne / 项目主管：Yi Eui-sung / 项目经理：Lim Sung-bum / 项目建筑师：Jon Ji-young, Lim Sung-soo, Zach Pauls, Aaron Ragan / 项目设计师：Daniel Pruske, Natalia Traverso-Caruana
项目团队：Ilaria Campi, Her Yoon, Kim Meari, Sarah Kott, Lee Michelle, Park Jung-jae, Seo Go-woon, Pablo Zunzunegui / 先进技术：Cory Brugger, Kerenza Harris, Stan Su, Atsushi Sugiuchi / 项目助理：Natalie Abbott, Viola Ago, Lily Bakhshi, Paul Cambon, Jessica Chang, Tom Day, Kabalan Fares, Stuart Franks, Fredy Gomez, Marie Goodstein, Parham Hakimi, Maria Herrero, James Janke, Kim Dongil, Lee One-jea, Lee Seo-joo, Katie MacDonald, Eric Meyer, Nicole Meyer, Elizabeth Miller, Liana Nourafshan, Brian Richter, Ahmed Shokir, Ari Sogin, Colton Stevenson, Henry Svendsen, Derrick Whitmire, Helena Yun, Eda Yetim / 本地建筑师：Haeahn Architecture / 结构：Buro Happold, SSEN / 机电：Arup, HiMec, Nara / 可持续性、LEED：Arup, Transsolar, HiMec, Eco-Lead / 立面：Arup, FACO / 照明：Horton Lees Brogden Lighting Design, Alto Lighting / IM：Morphosis, Gehry Technologies, DTCON Architecture / 景观：Morphosis Architects, Haeahn Architecture / 室内：Morphosis Architects, Haeahn Architecture, Kidea / 施工管理：Kolon Global Corp. / 总承包商：Kolon Global Corp. / 立面施工：Korea Carbon (GFRP), Korea Tech-Wall (GFRC), Han Glass (Curtain Wall), Steel Life (Interior Liner) / 客户：Kolon Industries, Inc. / 功能：corporate headquarters, offices, and research center including labs, meeting suites, exhibition space, brand shop, cafeteria, library, lecture rooms, and other support facilities / 用地面积：18,503m² / 建筑面积：76,300m²-phase 1; 22,503m²-phase 2 / 总建筑面积：98,803m² / 设计时间：2013.11—2015.11 / 施工时间：2015.6—2018.3 / 摄影师：©Jasmine Park (courtesy of the architect)-p.90~91, p.94, p.97, p.102, p.103, p.104, p.108, p.109; ©Kim Jong-oh-p.99, p.105, p.106~107

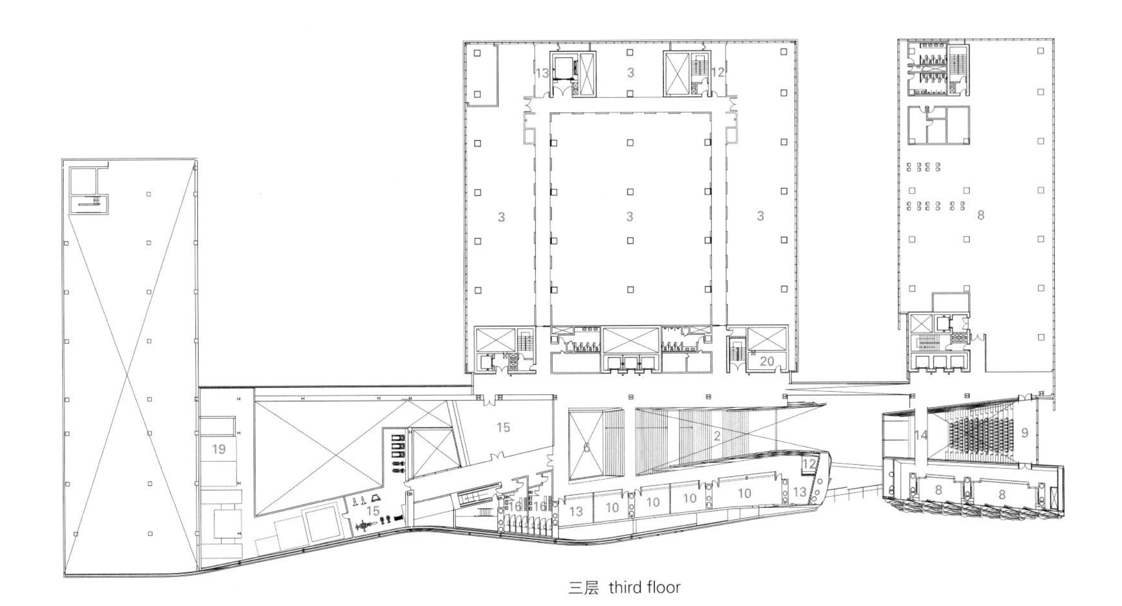

三层 third floor

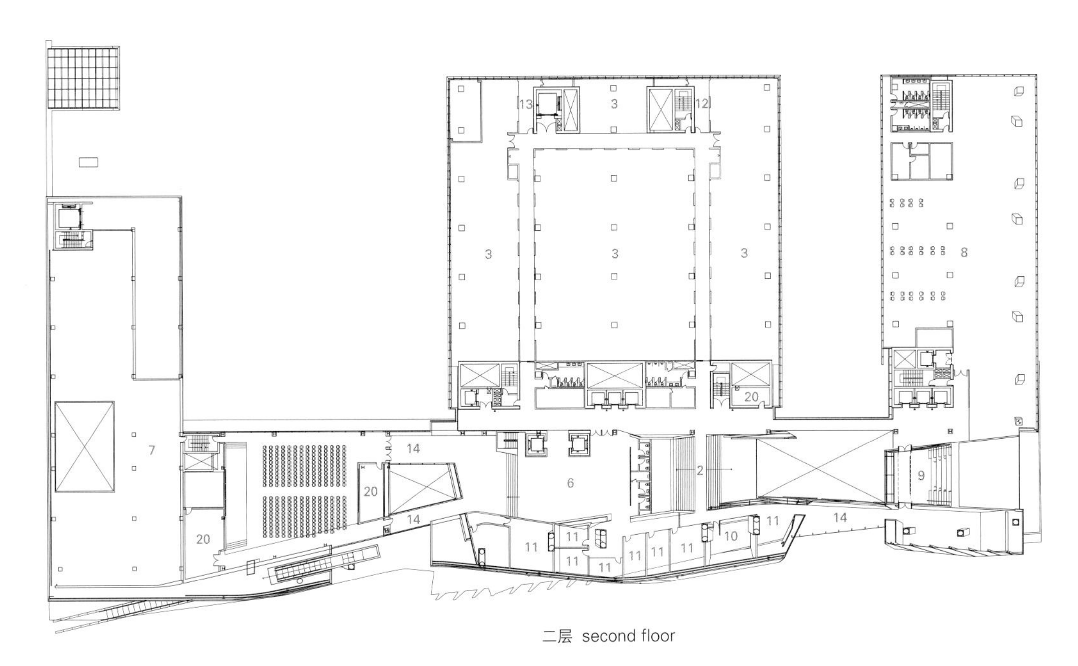

二层 second floor

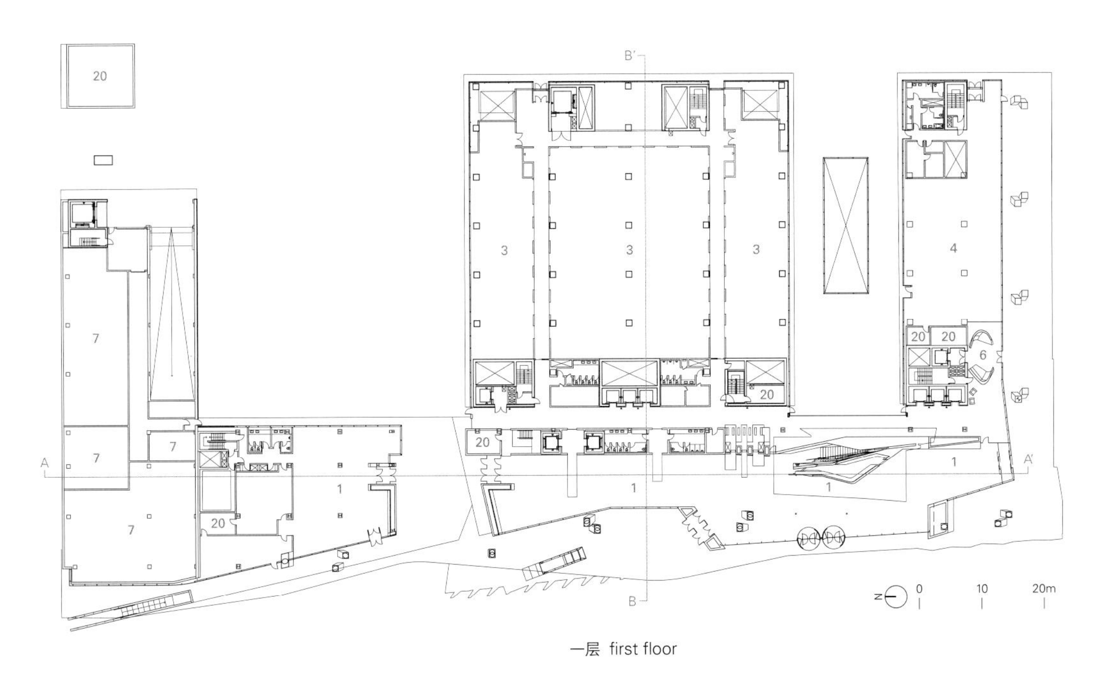

一层 first floor

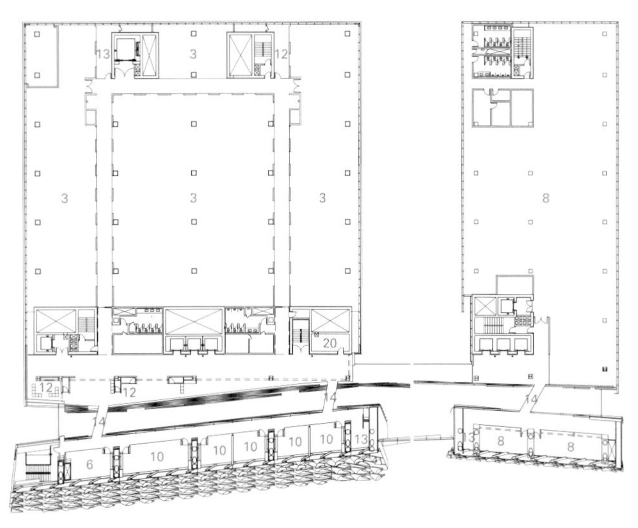

七层 seventh floor

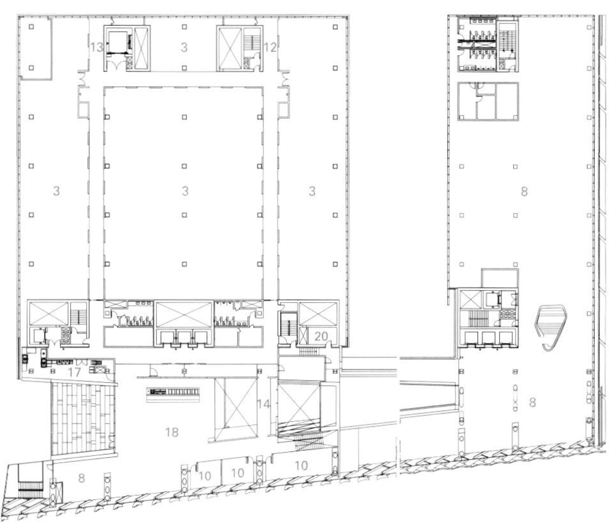

九层 ninth floor

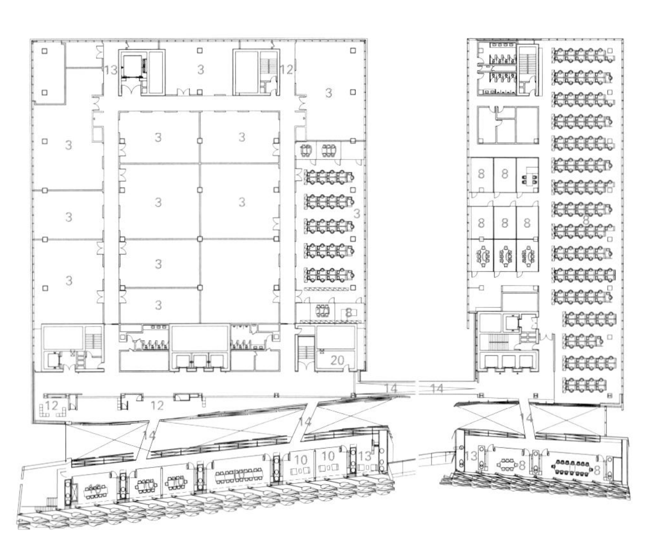

六层 sixth floor

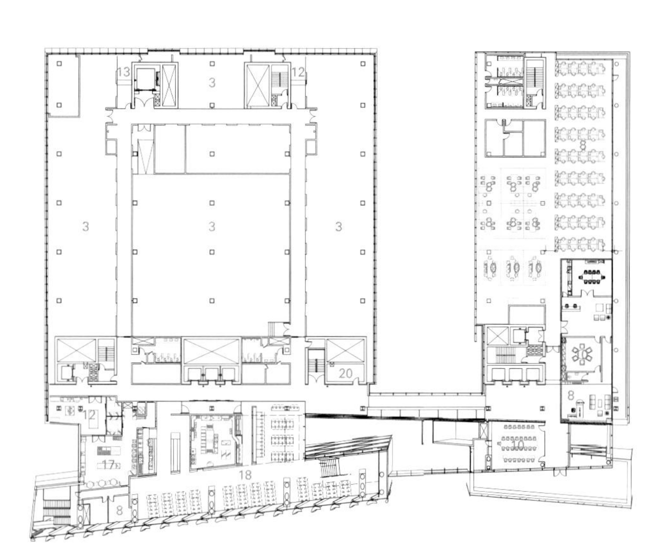

八层 eighth floor

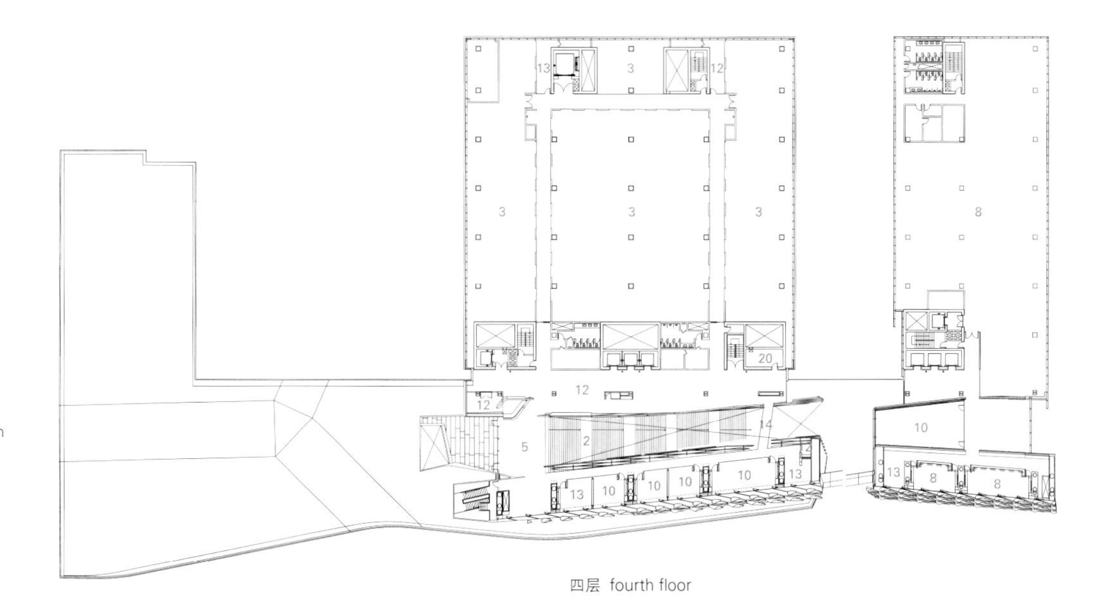

四层 fourth floor

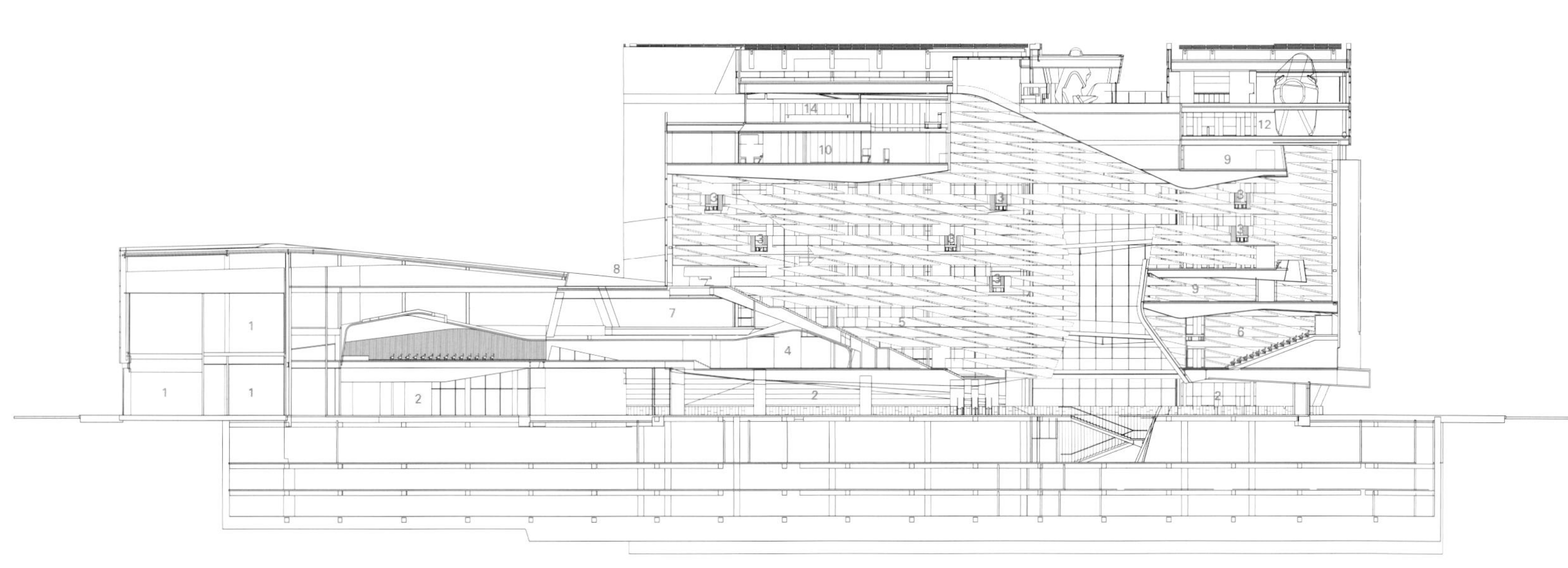

A–A' 剖面图 section A-A'

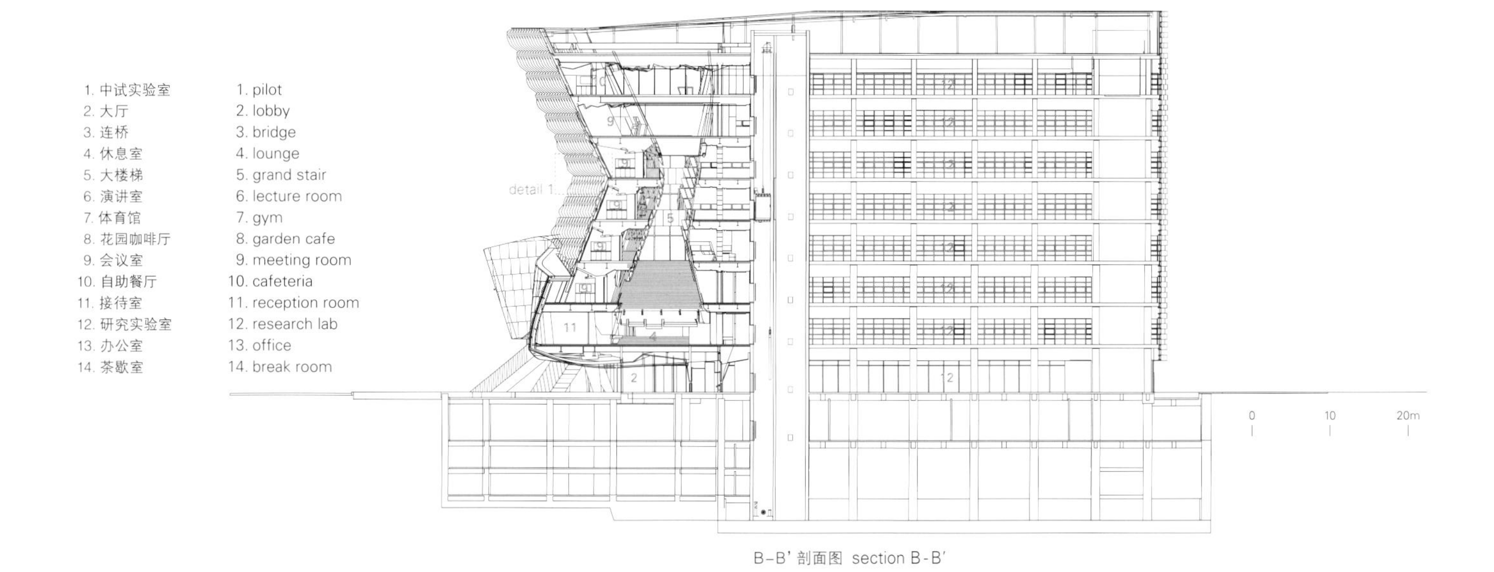

B-B' 剖面图 section B-B'

爱茉莉太平洋集团总部
Amorepacific Headquarters

David Chipperfield Architects

DRAGON HILL

韩国最大的化妆品公司爱茉莉太平洋集团的新总部位于首尔市中心，该公司自1956年以来一直占据着这个位置。新建筑位于首尔龙山区，位于一个正在开发的宽敞公园和一个高楼大厦林立的商业区之间。创建一座独具特色的建筑是该项目的主要目标。必须在两种截然不同的城市条件之间进行调解，从而强调了这一设计决定：同时，历史街区具有小规模的建筑层次结构，具有固有的城市品质，其总体规划显然受到了早期现代运动愿景的启发，在早期现代运动中，孤立的建筑通过形成一个由物体组成的城市来定义城市的特性。这种规模的建筑有一种公共责任，需要通过提供外观卓越的公共场所来实现这个目标，也是首尔所有公民都能享受城市生活活力的"去处"。对互联互通的渴望——内部和外部世界的自由沟通、开放互动、动态共存——源自一种深刻的人文主义企业精神，远远超出了总部的职责范围。这种设计哲学体现在一个致力于"形式服从目的"的整体方法的建筑概念中。

这个立方体体量共30层，总建筑面积21.6万平方米。中心内庭院和立面上的三个大洞口，每一个都包含一个花园，打破了紧凑的形式，使建筑渗透到城市，并允许空气和阳光渗透到它的深处。三个面向城市开放的大型洞口将这个中心空间与外部环境连接起来，让人能欣赏到城市和远处山脉的美景，因此建立了一种方向感和归属感。

作为"空中花园"，这些洞口呈现了建筑的规模，将大自然的美景从邻近的公园延伸到建筑的所有部分。

从一开始，新总部大楼的设计就受到这样一种理念的驱动：我们城市的质量不仅取决于单个建筑，还取决于它们对城市公共空间和社会空间的贡献。这座建成建筑在同一个屋顶下结合了高效的公司总部和各种公共设施，包括一个美术馆、一个大礼堂、一个图书馆、一个托儿设施以及餐厅、酒吧和咖啡馆，为不断发展的首尔大都市的市民提供了重要的新空间。

"它不仅仅是一个办公室，"大卫·奇普菲尔德建筑师事务所的合伙人兼创始人大卫·奇普菲尔德说，"这座建筑向在这里工作的人和市民展示了慷慨的精神。它是公司和城市之间的媒介。它展示了一家公司如何参与到更大的社区中去。"

这些社会文化价值也决定了建筑的形式。大楼的解决方案被否决了，取而代之的是一个宽和高约为100m的立方体。以这种形式创造的巨大空间通过与街面增加接触，为公共活动提供了更多的选择。

该建筑的所有建筑、结构和技术概念都源于广泛和全面的可持续设计方法。该设计的灵感来自韩国传统以及当地数百年来在气候、资源和工艺方面发展起来的专业知识。

N
0
100
200m
N
0
50
100m

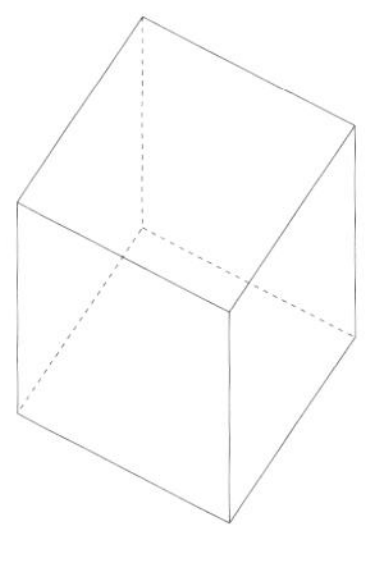

立方体 cube

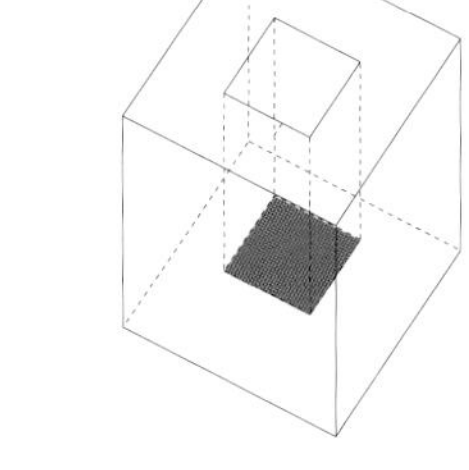

庭院 courtyard

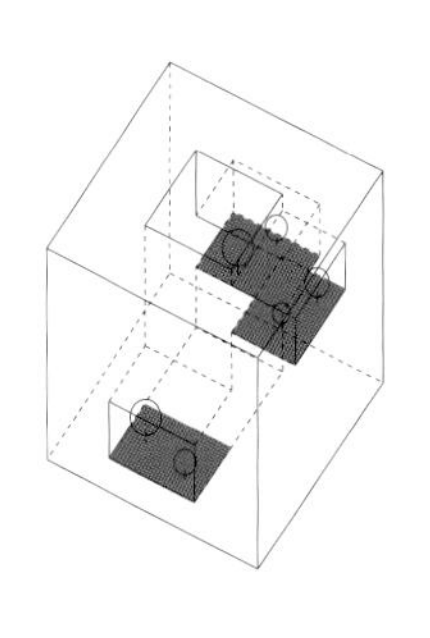

高架花园
elevated gardens

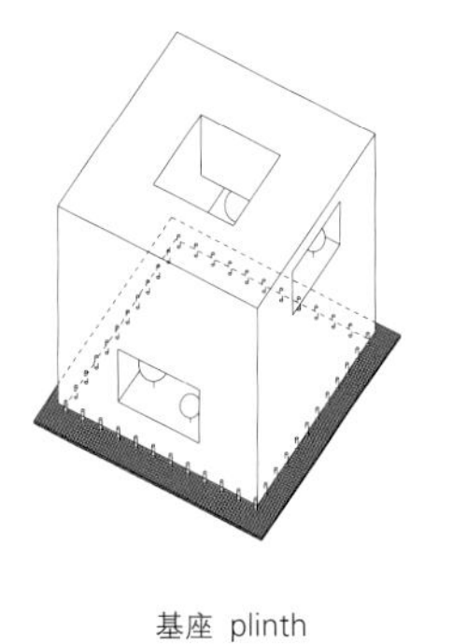

基座 plinth

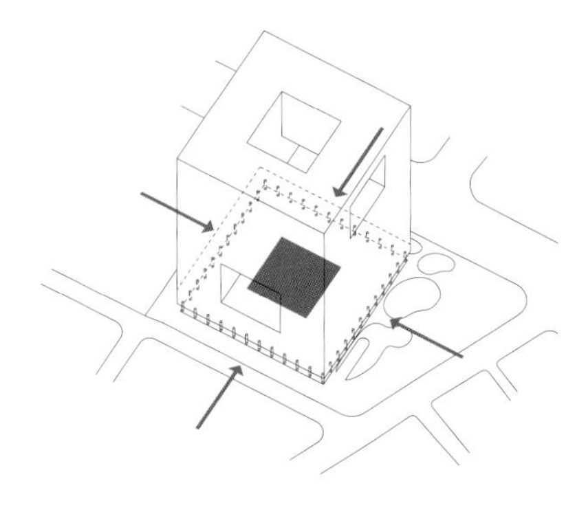

中央大厅
central lobby

外层包裹着建筑体量，它就像一层轻盈的、几乎为全纺织材料的窗帘，增加了外观的深度和有趣的细节。它由亚光白色的铝翅片组成。翅片的大小和形状经过精心调整，可以提供最佳的自然采光和通畅的视野，同时也减少了不必要的太阳辐射和室内眩光。因此，这些翅片具有不同的大小，并根据每个主要方向的太阳照射量分为四个不同的“家族”。

连通性成为室内办公室设计的主导理念。为了提供一个特别适合交流的、灵活的工作环境，团队区域和个人工作空间形成了一个相互连接的景观，以多种方式促进开放和非正式的沟通。室内楼梯与相关的服务中心连接着所有的办公室楼层。办公空间为会议和工作提供了各种各样的机会，而两者之间的界限仍然是流动的。因此得到了一个创新的工作环境，减少了层级制度，有利于平等和集体结构，通过在传统与进步、自然与人工、内在健康与外在美之间的平衡，呼应了公司的价值观。

The new headquarters for Amorepacific, Korea's largest beauty company, is located in the center of Seoul at a site which has been occupied by the company since 1956. The new building is situated in the district of Yongsan-gu, between the site of a spacious public park currently under development, and a business district of high-rise towers. Creating a building with a distinct identity was the primary aim for the project. This decision was underlined by the need to mediate between two distinctly different urban conditions: Whilst the historic neighbourhood features a small-scale architectural hierarchy with an inherent urban quality, the masterplan is clearly inspired by visions of the early

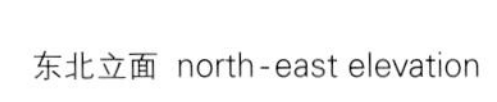

东北立面 north-east elevation

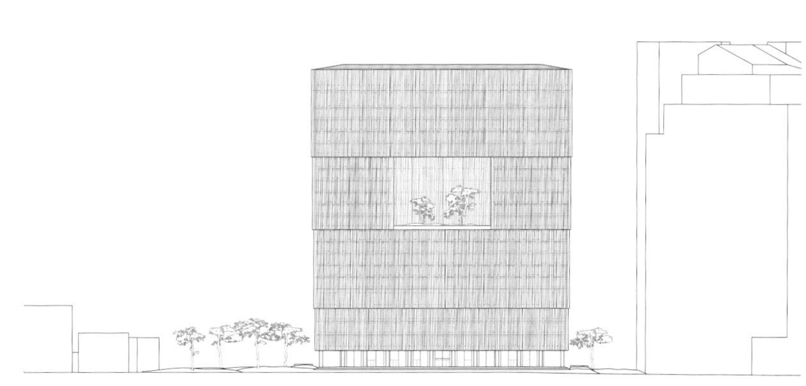

西北立面 north-west elevation

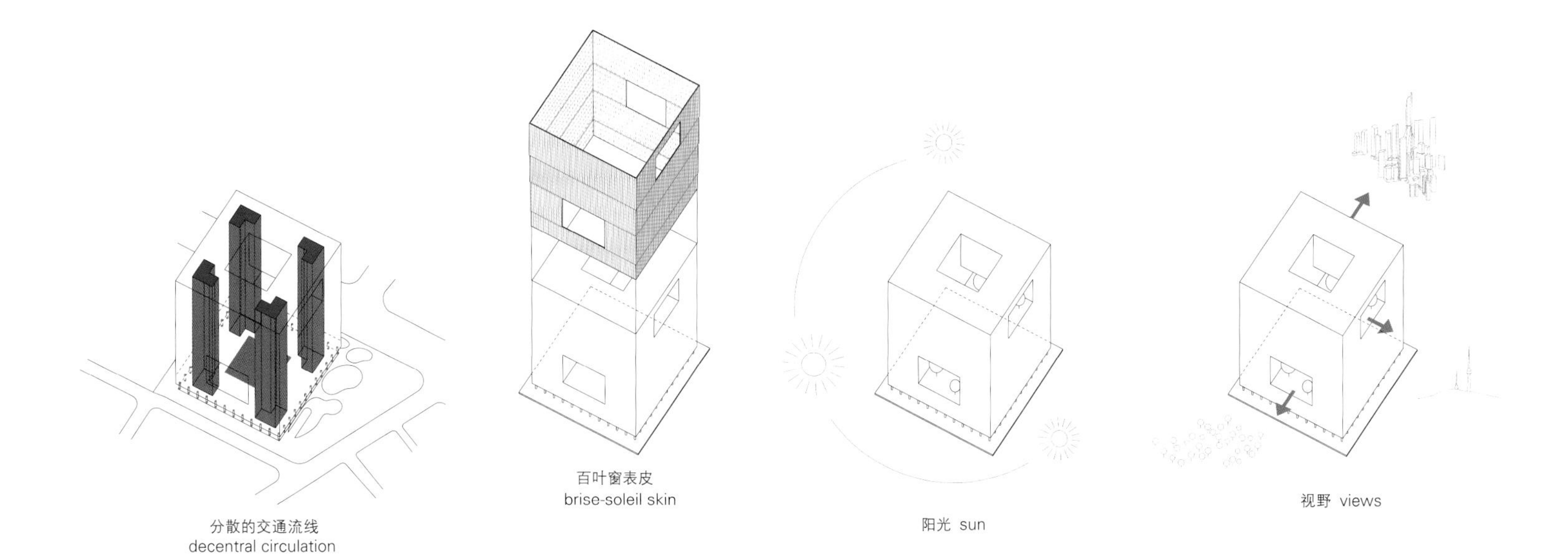

modern movement, where solitary buildings define urbanity by forming a city of objects. Buildings of this scale have a public responsibility that needs to be addressed by offering a public destination beyond formal appearance, a "place to be" where the dynamics of urban life can be enjoyed by all citizens of Seoul. The desire for connectivity – free communication, open interaction, and dynamic coexistence, both internally and with the outside world - arises from a profound humanistic corporate ethos that reaches far beyond the responsibilities of a headquarters. This philosophy manifests itself in an architectural concept that is dedicated to the integral approach of "form follows purpose".

The cube-shaped volume contains a total of 30 floors and a gross floor area of 216,000m². A central inner courtyard and three large openings in the facade, each containing a garden, break up the compact form, making the building permeable to the city and allowing air and daylight to filter into its depths. Three large urban openings connect this central void with the exterior surroundings, providing views over the city and the mountains in the distance and therefore establishing a sense of orientation and belonging. As "hanging gardens", these openings give scale and allow nature to extend from the adjacent park into all parts of the building.

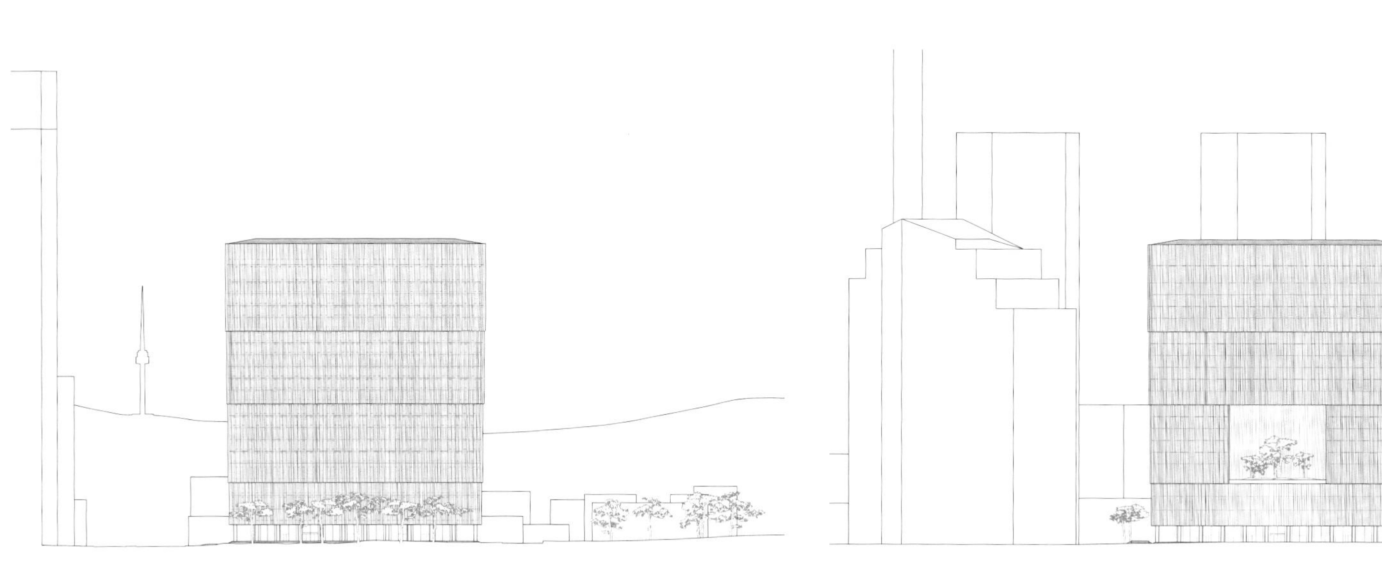

西南立面 south-west elevation

东南立面 south-east elevation

포차촌
PARIS BAGUETTE

한강로 전일제
중앙차로버스전용
부동산

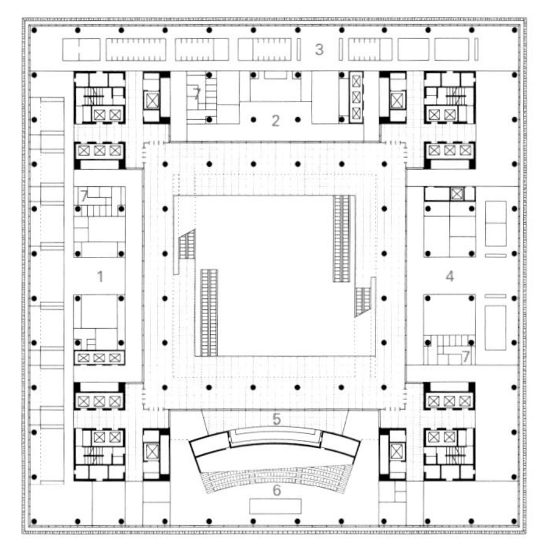

二层 first floor

1. 日托中心
2. 公司历史空间
3. 顾客测试区
4. 皮肤和头皮管理中心
5. 礼堂酒吧
6. 礼堂
7. 休息室

1. daycare center
2. company history space
3. test customer area
4. skin and scalp center
5. auditorium bar
6. auditorium
7. restroom

一层 ground floor

1. 中庭
2. 总服务台
3. 爱茉莉太平洋集团美术馆，大厅
4. 展览空间
5. 爱茉莉太平洋集团美术馆，图书馆
6. 茶室

1. atrium
2. main desk
3. Amorepacific Museum of Art, lobby
4. exhibition space
5. Amorepacific Museum of Art, library
6. tea room

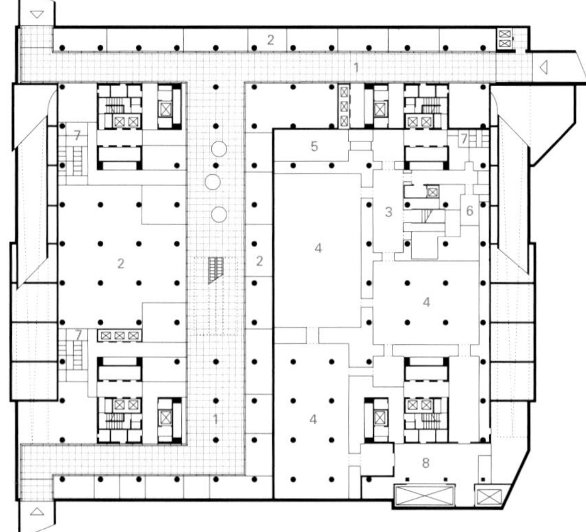

地下一层
first floor below ground

1. 地下通道
2. 零售和餐饮空间
3. 爱茉莉太平洋集团美术馆，地下大厅
4. 展览空间
5. 研讨室
6. 衣帽间
7. 休息室
8. 艺术配送处

1. underground passage
2. retail, food & beverage
3. Amorepacific Museum of Art, lower lobby
4. exhibition space
5. seminar room
6. cloakroom
7. restroom
8. art delivery

1. 开放设计的办公室
2. 休息室

1. open plan office
2. restroom

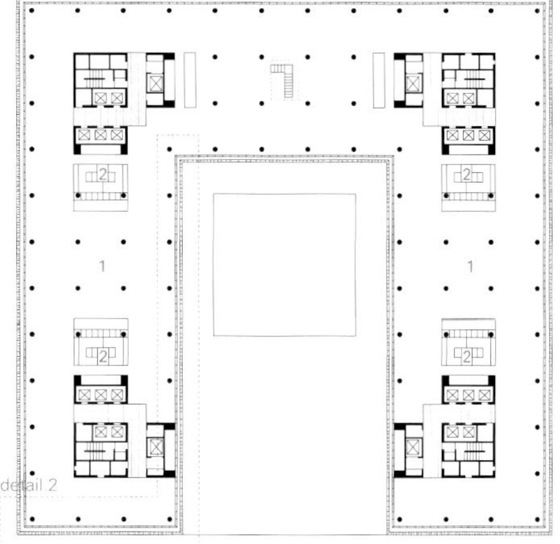

六层 fifth floor

1. 开放设计的办公室
2. 屋顶花园
3. 主服务中心
4. 休息室

1. open plan office
2. roof garden
3. main hub
4. restroom

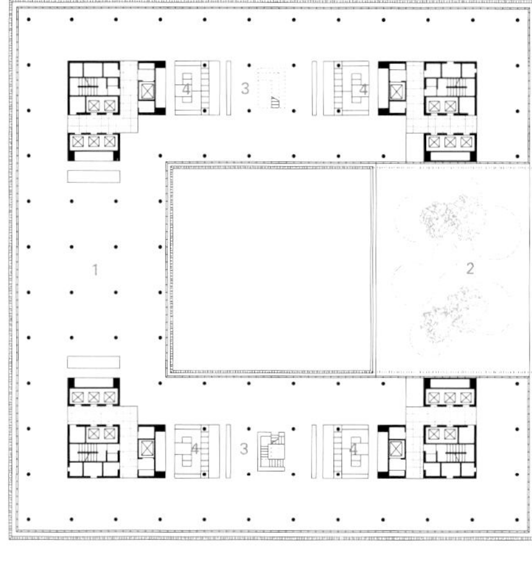

17层 sixteenth floor

1. 爱茉莉太平洋集团餐厅
2. 咖啡馆
3. 健身房
4. 康复中心
5. 屋顶花园
6. 休息室

1. Amorepacific restaurant
2. cafe
3. fitness
4. healing center
5. roof garden
6. restroom

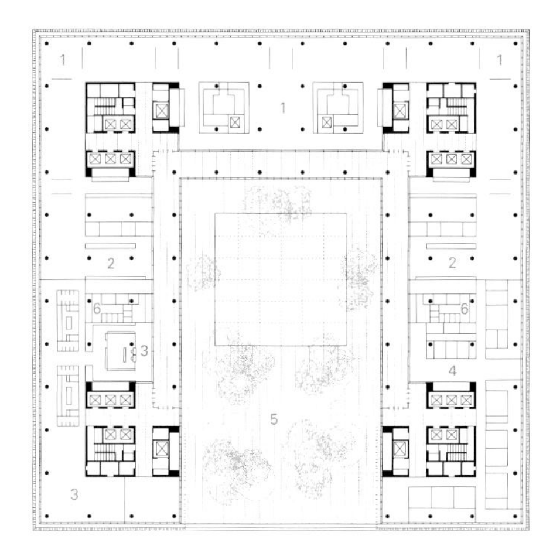

五层 fourth floor

1. 开放设计的办公室
2. 屋顶花园
3. 主服务中心
4. 休息室

1. open plan office
2. roof garden
3. main hub
4. restroom

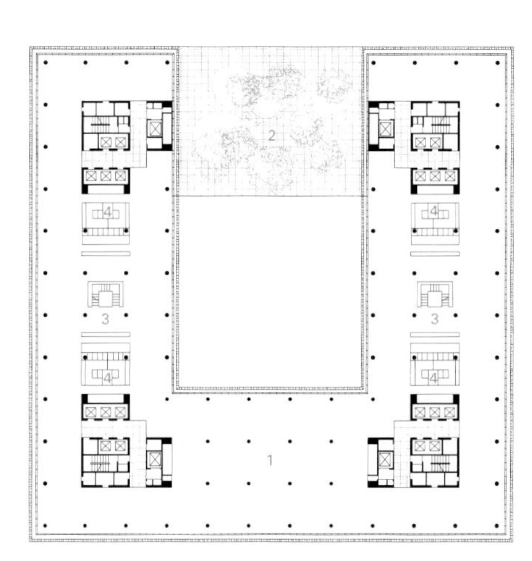

11层 tenth floor

1. 礼堂
2. 爱茉莉太平洋集团美术馆，办公室
3. 会议室
4. 会议中心
5. 休息室

1. auditorium
2. Amorepacific Museum of Art, office
3. conference room
4. conference hub
5. restroom

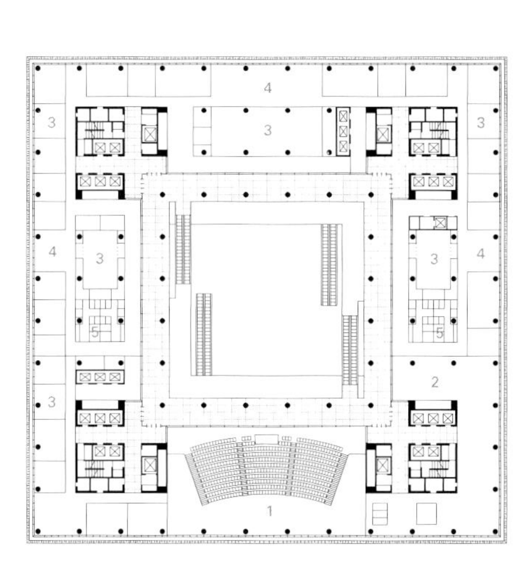

三层 second floor

1. 开放设计的办公室
2. 会议室
3. 主服务中心
4. 次服务中心
5. 休息室

1. open plan office
2. meeting room
3. main hub
4. secondary hub
5. restroom

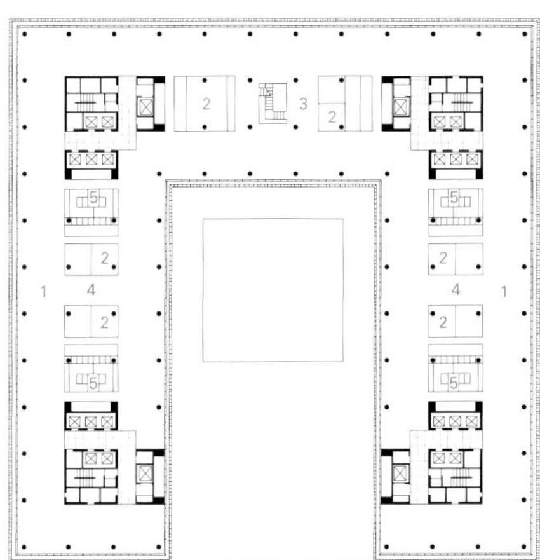

八层 seventh floor

From the very start, the design for the new headquarters building was driven by the notion that the quality of our cities does not depend on individual buildings alone, but on the contribution they make to the common and social space of the city. The completed building combines under one roof an efficient company headquarters together with diverse public facilities, including a museum, a large auditorium, a library, a childcare facility as well as restaurants, bars and cafés – providing significant new spaces for the citizens of the growing metropolitan city of Seoul.

"It is more than an office," states David Chipperfield, Partner and founder of David Chipperfield Architects. "The building suggests generosity of spirit to the people who work here and the citizens. It is something that mediates between the company and the city. It shows how a company can participate in the larger community."

These social and cultural values also determined the architectural form. A tower solution was rejected in favour of a cube of approximately 100 m in width and height. The large footprint created by this form resolutely provides more options for communal activities through greater engagement with the street level.

All architectural, structural and technical concepts for the building stem from a broad and holistic approach towards sustainability. The design is inspired by Korean traditions as well as local expertise that have been developed over centuries in terms of climate, resources and craftsmanship.

The external layer envelopes the volume like a lightweight, almost textile curtain, adding depth and playful detail to the external appearance. It consists of matt-white aluminium fins. Size and shape of the fins have been carefully tuned to provide optimal natural lighting and unobstructed views, while reducing unwanted solar radiation and glare for the interiors. The fins therefore have different sizes and are grouped in four different "families" as dictated by the solar exposure for each cardinal direction.

Connectivity became the leading idea for the interior office design. In order to provide a highly communicative and flexible working environment, team zones and individual workspaces form an interconnected landscape that facilitates open and informal communication in diverse ways. Internal staircases with associated hubs connect all office floors. The office space provides various opportunities for meeting and working, while the boundaries in between remain fluid. The result is an innovative working environment that reduces hierarchies in favour of egalitarian and collective structures, echoing the company values by oscillating between tradition and progress, naturalness and artificiality, inner health and outer beauty.

OPENING on 20 8 January 2nd !
New Year

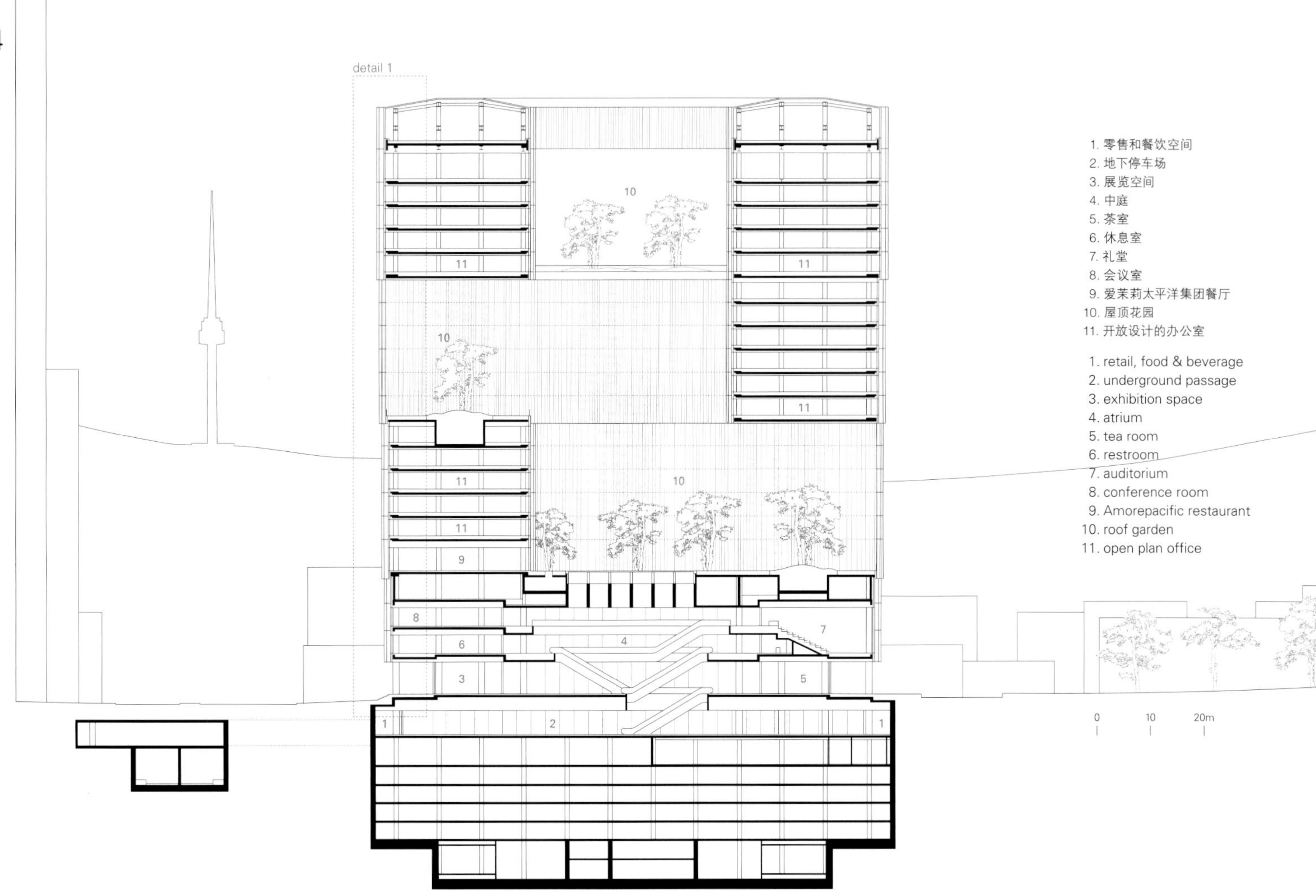

A-A' 剖面图 section A-A'

B-B' 剖面图 section B-B'

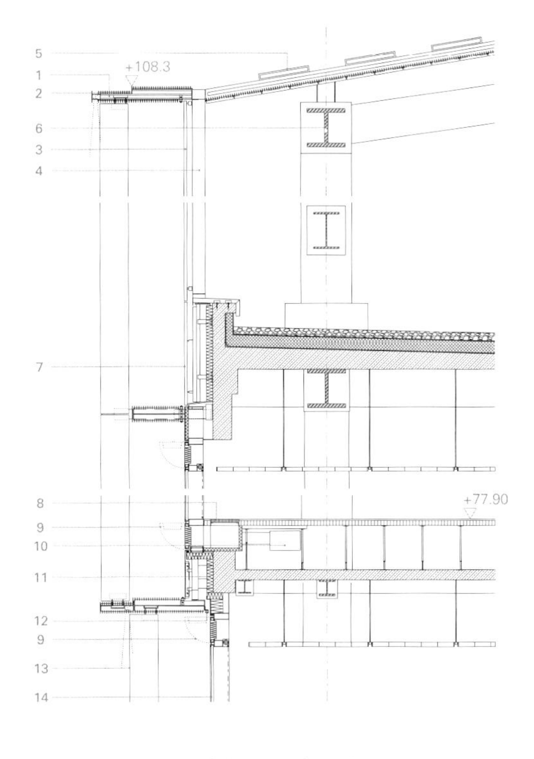

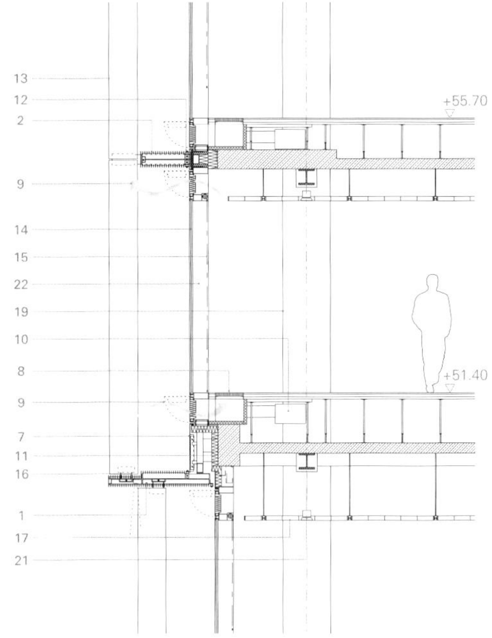

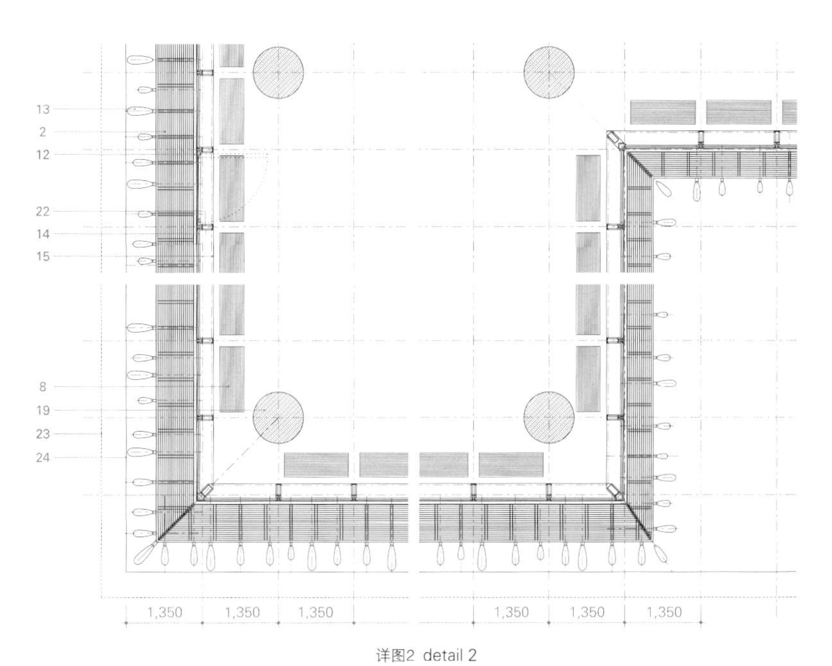

详图2 detail 2

+22.70

+7.50

18

19

20

-1.20

详图1 detail 1

1. steel construction
2. horizontal catwalk for shading and maintenance, prefabricated aluminium grille, coated
3. vertical aluminium grille, coated
4. aluminium extrusion profile, sub construction for vertical grille, coated
5. photovoltaic collectors
6. steel bridge construction
7. aluminium cladding, coated
8. prefabricated aluminium grille, coated
9. motorised aluminium flaps for natural ventilation, insulated, coated
10. fan coil unit
11. insulation
12. latch rail for maintenance
13. vertical aluminium extrusions for shading, four sizes, coated
14. fixed triple glazing
15. internal glare protection, textile blind
16. facade lighting
17. suspended ceiling, open metal grille, aluminium, coated
18. reinforced concrete ceiling, fair-faced
19. concrete columns, fair-faced
20. solid block steps, granite, flamed and brushed
21. suspended lighting, part of bespoke luminaire family
22. prefabricated facade elements, coated aluminium frame, with access for maintenance
23. facade step, tenth floor
24. facade step, fourth floor

项目名称：Amorepacific Headquarters / 地点：Hangang-ro 100, Yongsan-gu, Seoul
建筑师：David Chipperfield Architects Berlin / 合伙人：David Chipperfield; Christoph Felger - Design lead; Harald Müller
项目建筑师：Hans Krause - Overall project management; Nicolas Kulemeyer and Thomas Pyschny - Project management (Construction documentation, Construction administration phase) / 合作方：HAEAHN Architecture (Seoul) - Shell & Core, Schematic design + Construction documents; KESSON (Seoul) - Interiors
工地监理：Kunwon Engineering Co. Ltd. (Seoul) / 总承包商：Hyundai Engineering & Construction (Seoul) / 设计工程师：Arup Deutschland GmbH (Berlin) + Arup Ltd. (London) - Schematic design~Design development / Signage: L2M3 communication design (Stuttgart) / 景观设计师：SeoAhn (Seoul) / 客户：Amorepacific Corporation / 用途：offices, gastronomy, fitness, library, daycare center, company history space, test customer area - Amorepacific facilities; atrium and exhibition space, Amorepacific Museum of Art, Amorepacific Museum of Art library, auditorium, conference center, gastronomy, tea room, retail including company brands - Public facilities; tenant offices - Lettable space; pocket park, roof gardens - Gardens / 平均海平面高度：±0.00m = 12,80m / 楼高：110m
用地面积：14,500m² / 建筑面积：8,700m² / 总建筑面积：216,000m² / 长和宽：approx. 5,800m² above ground, approx. 9,800m² below ground / 长和宽：90m x 90m (ground floor) / 楼层：23 stories above ground (Ground floor + 22 floors), 7 stories below ground / 最大容量：approx. 7,000 staff plus pedestrian traffic
礼堂：450 seats / 停车：680 spaces / 可持续性标准：LEED Gold (certification expected for summer 2018) / 项目开始时间：2010 / 施工时间：2014—2017
摄影师：©Noshe (courtesy of the architect) - p.112, p.116~117, p.121, p.122~123; ©Amorepacific Corporation - p.110~111, p.125; ©Hyun Yu-mi (C3) - p.120[right]; courtesy of the architect - p.120[left], p.126, p.127

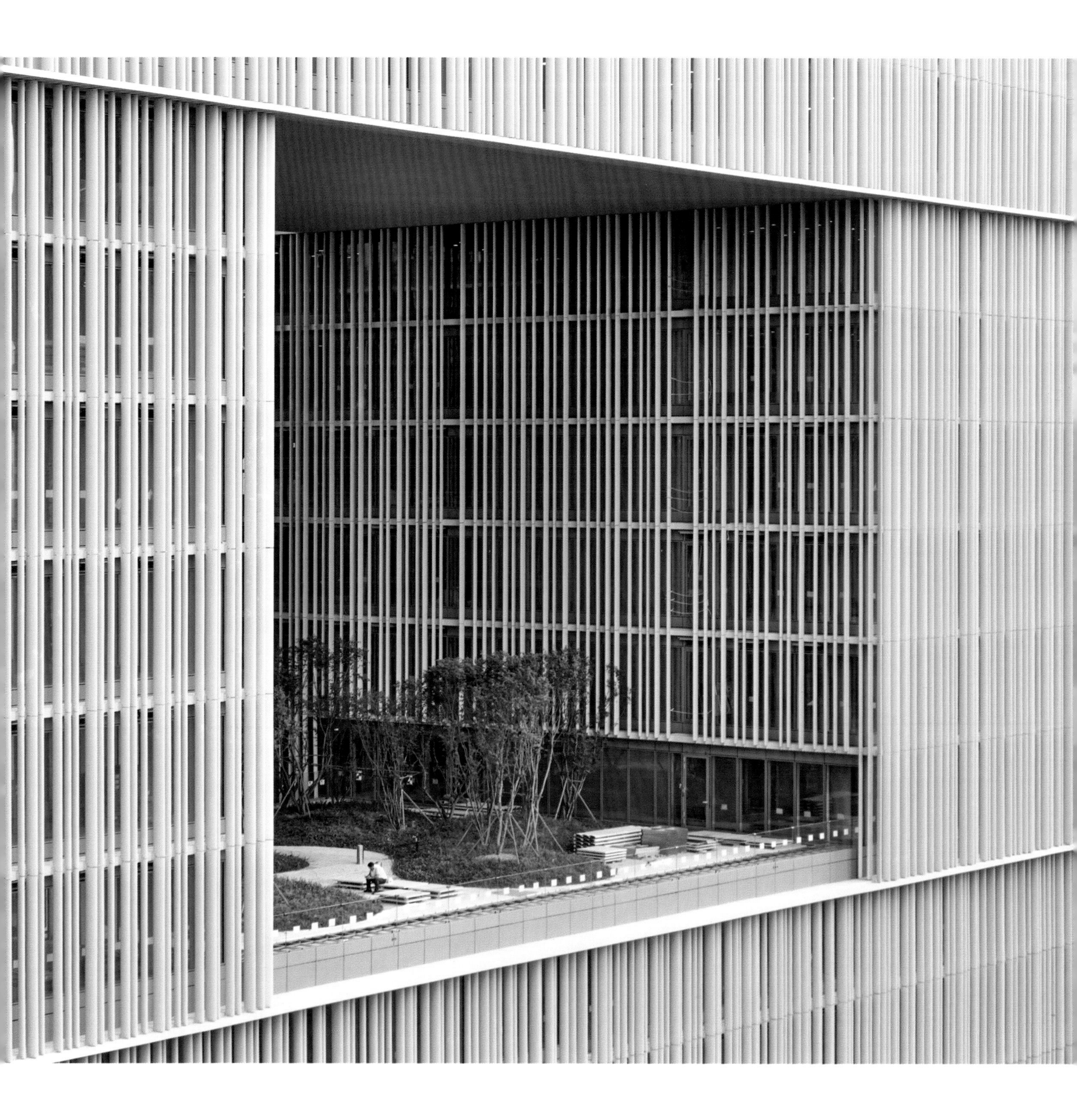

品牌与连通性

大卫·奇普菲尔德访谈，2018年6月14日，首尔爱茉莉太平洋集团总部

Brand and Connectivity

Interview with David Chipperfield, 14 June 2018, at the Amorepacific HQ., Seoul

玄瑜美（以下简称“C3”）：爱茉莉太平洋C3公司总部位于龙山区，这里有着重要的历史、城市和社会问题。它们是如何影响您的设计的？

大卫·奇普菲尔德（以下简称“DC”）：在早期阶段，建筑的大环境非常重要，但由于有大量的建设正在进行，所以我并不清楚。我的经验告诉我，现在很难在所有复杂的规划问题在现实生活中发生之前进行协调，因为我们都无法控制。然而，我们可以肯定的是，龙山区将变得更加密集，将会建起更多的高楼大厦。问题是，在这样的城市大环境之下，我们如何为城市做贡献？我相信，像这样的项目能做的就是激励未来可能发生的事情。我们把建筑本身当作一个入口。它必须是连接龙山公园、较高密度的城市或任何意想不到形式的加强点。我希望，不仅仅是建筑的基本结构，还有空间结构——人们的活动方式和社会空间的组织方式——都将是永恒和普遍的。

C3：简单的立方体形式是否也受到龙山区大环境的影响？

DC：这种嘈杂的环境鼓励我们不选择更复杂的形式，而是去开发非常简单的形式。我们相信，在这样嘈杂的大环境下，安静的建筑会给人留下更深刻的印象。但是它的形式不能太封闭，这样一座总建筑面积20万平方米的大建筑肯定会影响城市的景观。然后，我们开始思考如何让规模这么大的建筑舒适地坐落在城市里。所以，我们通过切割建筑的结构来打破简单的形式，并赋予它一系列的尺度。水平线、四个体量、窗户和翅片的设计都有助于在大型建筑中找到人性化的尺度。

身为外国建筑师，我们致力于如何建造属于韩国首尔的建筑。“月光罐”成了一种主题。我一直对韩国的陶瓷很感兴趣，月光罐是其中一个亮点。

C3：在这个项目中，你们似乎已经做出了很多努力来实现企业的社会责任。作为客户的爱茉莉太平洋集团是如何考虑这种方法的？

DC：虽然徐董事长非常热衷于建造一座原创建筑，但他更强调的是建筑本身能够体现公司的理念和原则。他有两个主要的担忧：第一个是工作场所的质量，第二个是建筑对城市的贡献。这座建筑应该是一个鼓励员工参与的工作场所。更重要的是，它应该是一个为在大楼里工作和生活在城市里的人提供社交活动的场所。因此，该建筑配备了完善的社交设施，如咖啡馆、美术馆、图书馆、幼儿园和礼堂。因此，我们开始使用可以创造这种公共空间的立方形式的概念，为工作环境提供更多的可能性。

Yumi Hyun, C3 *Amorepacific Headquarters sits in Yongsan where there are crucial historical, urban and social issues. How did they affect your design?*

David Chipperfield At the early stage, the context of the building was very important yet unclear as there was an enormous amount of construction going on. My experience tells me that nowadays it is very difficult to coordinate all the complicated planning issues before they happen in real life, as none of us have control over it. Yet, we knew for sure that Yongsan will become denser and there will be more towers. The question was – in such urban context, how can we contribute to the city? I do believe what projects like this can do is to stimulate things that might happen in the future. We treated the building itself as a gateway. It has to be the enhanced point of connection to the Yongsan park, the city with higher density, or any unexpected forms. Hopefully, not only the basic structure that holds the building, but the spatial structure – how people move around and how social spaces are organized, will be timeless and universal.

Yumi Hyun *Is the simple cubic form also influenced by the context of Yongsan?*

David Chipperfield This noisy context encouraged us to develop from more complicated forms to a very simple form. We believed that in this noisy context, the quiet building would make more impression. But its form must not be so enclosed, as such a big building with gross floor area of 200,000m² would definitely affect the urban landscape. Then we started to wonder how to make such a big building sit comfortably in the city. So, we cut the building by its structure to break from a simple form and gave it a range of scales. The horizontal lines, the four volumes, the windows and the fins all contribute to finding human-scale in a large building.

We, as foreign architects, addressed ourselves how to make the building that belongs to Seoul, Korea. The "moon jar" became a sort of motif. I've always been interested in Korean ceramic and the moon jar is one of the high points.

Yumi Hyun *It seems that many efforts have been made to materialize corporate social responsibility in this project. How did the client, Amorepacific, consider such approach?*

David Chipperfield Though Chairman Seo was very keen to build an original building, he emphasized more on the building that would embody ideas and principles of the company. He had two major concerns: the first was the quality of a workplace, and the second was the contribution that the building can make to the city. The building should be a workplace that encourages the engagement of the office workers. More importantly, it should be a social place for people who work in the building and people who live in the city. Thus, the building is highly equipped with social facilities like cafes, museum, library, kindergarten, and auditorium. Therefore, we started to work

我同意，建筑设计是硬件，而使用它的人是软件。我们在建筑中所能做的就是创建一个好的框架，而在这个框架中软件会做得更好。因此，我们想确保硬件足够灵活，以适应社会的变化。

C3：与其他办公楼不同的是，爱茉莉太平洋集团总部大楼的四侧立面都有入口，这些入口似乎具有同等的重要性。

DC：从面向城市、河流、山脉、公园的四扇门进入。这座建筑坐落在城市中，与传统的办公大楼截然不同。通过在建筑中引入美术馆、咖啡馆、商店等设施，我们可以将城市景观带入其中。这样，城市和建筑之间形成了一种"柔性的融合"。

C3：中庭和巨大的屋顶花园在实际上和概念上都是开阔的。您设计成这么开放的意图是什么，尤其是在公司大楼里？

DC：公共空间带来了社区的感觉，并将建筑变成一个小村庄，以简单的形式保持复杂性。较低的楼层成为中庭，这是大楼里接待和分配人员的关键空间。它确实是上班族和城市行人的交汇点。同样，屋顶花园强化了社会空间的重要意义。这些公共空间实现了这个想法：不同寻常的、不实用的空间也能带来更多的体验，成为适合的"场所"。在每一次尝试中，它的目标都是履行自己的社会责任，这将项目扩展到了一个更广泛的概念，即公司如何在城市中发挥作用。

C3：白天，建筑看起来很坚固，但到了晚上，透过密集的铝翅片，柔和的光线使建筑变得模糊和半透明。您的立面设计策略是什么？

DC：当然，这座建筑给人的感觉是坚固而透明的。这是一座玻璃建筑，为办公区域带来尽可能多的光线，同时通过铝翅片保护室内免受阳光直射，避免室内过热。通过在不同角度精心布置四种不同尺寸的翅片，我们能够在不妨碍朝向外部的视野的情况下，将充足的阳光引入室内。

这也有助于实现我们的团队和爱茉莉太平洋集团之间的共同目标，使建筑达到更高的环境标准。我们试图在技术和建筑设计方式上做出反应，为庭院和屋顶花园创造出巨大的空间。所有的被动式设备对采光和通风都很重要。例如，立面和洞口可以让尽可能多的自然阳光进入办公室，减少人工照明和能源消耗。

with the concept of cubic form that creates such common spaces giving more possibilities for a work community.
I agree that architecture is a hardware and people who are using it are a software. All we can do in architecture is to create a good framework, within which software will do better. Therefore we wanted to make sure the hardware is flexible enough to accommodate the changes in the society.

Yumi Hyun *Unlike other office buildings, Amorepacific Headquarters has entrances from all four sides that seem to claim equal significance.*
David Chipperfield You can enter the building from four gates that face city, river, mountain, and park respectively. The building sits in the city in a very different way than conventional office buildings. By introducing facilities like museum, cafes, and shops in the building, we could bring the urban landscape inside. In this way, the city and the building formed a 'soft integration' between each other.

Yumi Hyun *The atrium and huge roof gardens look wide and high open physically and conceptually. What was your intention with such openness, particularly in a corporate building?*
David Chipperfield The common spaces bring a sense of community and turn the building into a little village that holds complexity within a simple form. Lower levels of the building become the atrium, the key space that receives and distributes people in the building. It truly is the meeting point for the workers in the office and the pedestrians in the city. Similarly, the roof gardens enforce the matter meaning of social space. These common spaces realize the idea that, unusual and unfunctional space brings more experience and becomes "the place". In every attempt, it aimed to fulfill its social responsibility, which expanded the project to a wider idea of how the company plays a role in the city.

Yumi Hyun *In the day time, the building looks quite solid, but it turns ambiguous and translucent at night by the soft light through the dense aluminum fins. What was your strategy with the facade design?*
David Chipperfield Certainly, the building feels solid yet transparent. It is a glass building which brings as much light into the office areas, yet protects the interior from heat and direct sunlight by aluminum fins. By placing fins with four different sizes carefully at different angles, we were able to invite adequate daylight to the inside without obstructing the view towards the outside.
It also helps to achieve the common aim between our team and Amorepacific to make a building with high environmental standards. We tried to react in both technical and architectural ways, creating big voids for courtyard and roof gardens. All the passive devices are important for light and ventilation. For instance, the facade and the openings allow as much natural sunlight in the office to reduce the artificial lighting and energy consumption. Summarized by Boo Eun-bin

Kiswire釜山总部 + F1963

Kiswire Busan Headquarters + F1963

BCHO Architects

与土地的关系

一段时间以来，韩国电线制造商Kiswire公司一直计划在釜山附近的Mangmidong开发该公司名下的一块场地。从2014年开始建设宿舍和博物馆，到2016年又新建了公司总部。这些重建项目的组合包括名为F1963的文化综合体，它本身于2016年8月完成。

Kiswire公司总部位于建筑群其他设施之间的一座山上，与周围环境的高度相差25m。为了最大限度地减少对现有山丘的破坏，办公楼的设计力求协调与旧工厂空间和相邻的公司博物馆之间的立面差异。

该建筑包括四层停车场、三层办公室以及与附近博物馆对齐的屋顶平台。由于建筑的上部体量被推入山体5m，所以办公室的整体组成似乎从地面上模仿了现有的山坡地形，试图通过视觉的连续性来保护博物馆和F1963之间的自然景观。

开敞布置及庭院

办公楼的体量为86m×20m，采用后张力结构，可以完全不使用梁柱。这样就留下了一个简单的空盒子。此外，它的设计通过将管道和电气托盘等机械系统隐藏在立面窗户和地板下方，节省了建筑预算和维护费用。外立面设计的目的是具有人性化的尺度，设置推拉窗，方便居民使用，也降低建筑的碳足迹，减少不必要的供暖和制冷需求。

悬挂结构避免了在办公空间内架设梁和柱，让这个巨大、开放的室内非常适合作为当代工作场所。建筑师将暖通空调系统移到主要空间之外，进一步净化了空间，在这里，建筑系统的唯一表现形式仍然是稀疏排列在光滑的混凝土天花板上的照明设施。因此，有限的材料和智能系统管理为办公室内部提供了视觉上的舒适。

该建筑既追求履行环境责任，也追求视觉舒适度。80m高的体量内的三个庭院创造了垂直堆叠的效果，使室内空间保持通风。景观屋顶位于建筑上方两米的位置，形成一个水平通道，这里可以通风，同时也能遮蔽建筑，使其免受阳光直射。尺寸适宜的室内窗户让整个办公空间在打开时可以实现自然通风。

沿着建筑东侧的一个大立面通过其铁丝网种植营造了一种生态友好的感觉，积极地反映了该公司的身份。因此，该项目既考虑了设计的可持续性，也考虑了对自然环境的影响，同时还充分解决了视觉舒适度和内在性的问题。

Relationship with the Land

Developing a former factory site to house a museum, office, dormitory and cultural complex, Korean wire manufacturer Kiswire had for some time planned to develop a site the company owned in the Busan neighborhood of Mangmi-dong. Beginning with the construction of a dormitory and museum in 2014, the company's headquarters was added in 2016. The combination of these individual redevelopments comprises the facility's cultural complex, called F1963, which itself was completed in August of 2016.

The Kiswire Headquarters itself is situated on a hill between the compound's other facilities, giving it a 25-meter difference in height from its surroundings. To minimize damage to the existing hill, the office building was designed to cooperate with this elevational difference between the old factory spaces and the adjacent company museum.

The building incorporates four stories of parking lots, three stories of offices and a roof deck aligned with the nearby museum. Since the structure's upper volume is pushed back into the hill 5 meters, the overall composition of the offices appear to mimic the existing topography of the hillside from ground level, attempting through visual continuity to preserve the natural landscape between the museum and F1963.

Wondong I.C.
Beonyeong-ro
Suyeong Tunnel
1
2
3
4
5
0
20
50m
1. 博物馆
2. 培训中心
3. 宿舍
4. 办公室
5. F1963
1. museum
2. training center
3. dormitory
4. office
5. F1963

©Kiswire (courtesy of the architect)

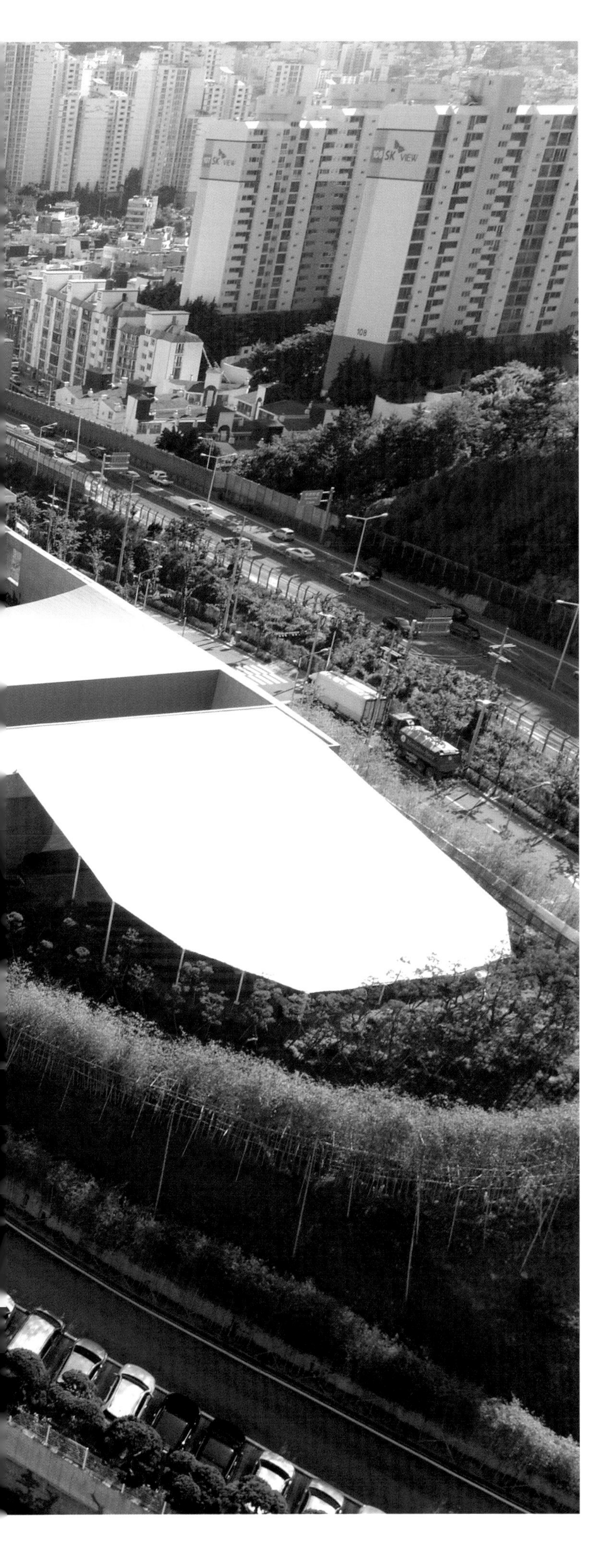

Open Plan and Courtyard

The mass of the office building, 86x20m, is constructed in a post-tension structure, allowing the complete exclusion of beams and columns. This leaves the space as a simple, empty box. Moreover, its design economically saves on the construction budget and maintenance by concealing mechanical systems such as ducts and electrical trays underneath facade windows and below the floors. The exterior facade design aims to entice a human scale, positing sliding windows which are both convenient for inhabitants and productive to lower the building's carbon footprint from unnecessary demands for heating and cooling.

The hung structure enables an absence of beams and columns in the office spaces, enabling a large, open interior well suited for contemporary workplaces. Relocating HVAC systems outside of the main spaces further clears the space, where the only expression of buildings systems remains thin, arrayed lighting on the smooth concrete ceiling. The interior of the office thus provides visual comfort with limited materials and intelligent systems management.

The building pursues both environmental responsibility and visual comfort. The three courtyards within the 80-meter volume create vertical stack effects to ventilate the interior spaces. The landscaped roof's location two meters above the building forms a horizontal passage through which wind is allowed to permeate while simultaneously shading the building from direct sunlight. Comfortably sized windows on the interior allow a capacity for the entirety of the office spaces to be naturally ventilated when opened.

A large facade along the building's eastern face positively reflects the company's identity by fostering a perception of eco-friendliness with its wire trellis plantings. As a result, the project arbitrates between sustainability and considerations of nature, while nonetheless adequately addressing questions of visual comfort and interiority.

用地面积：12,610m² / 建筑面积：2,078.61m² / 总建筑面积：13,049.82m²
建筑规模：three stories underground, four stories above ground
结构：post-tention, reinforced concrete
设计时间：2013.7—2014.3 / 施工时间：2014.6—2016.3 / 竣工时间：2016.8

kiswire

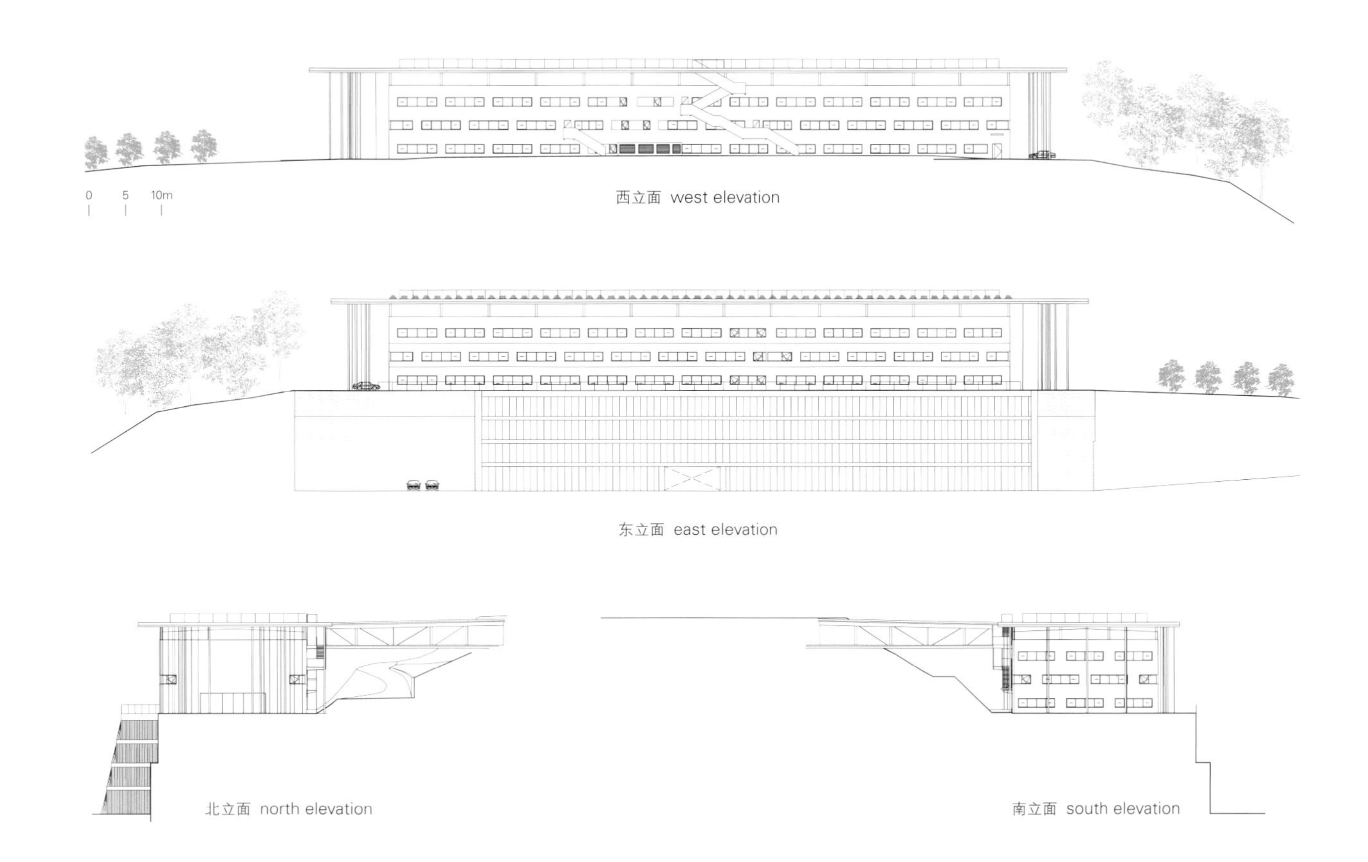

项目名称：Kiswire HQ, Busan / 地点：37, Gurak-ro 141beon-gil, Suyeong-gu, Busan, Korea / 建筑师：BCHO Architects / 项目团队：Kwon Do-yeon, Choi Ha-young, Park Ha-hyuck, Moon Han-sol, Shin Myung / 施工监理：Kwon Do-yeon, Moon Han-sol, Shin Myung / 施工方：Daewoo Development Co.,Ltd
客户：Kiswire / 用途：office / 摄影师：©Sergio Pirrone (courtesy of the architect) (except as noted)

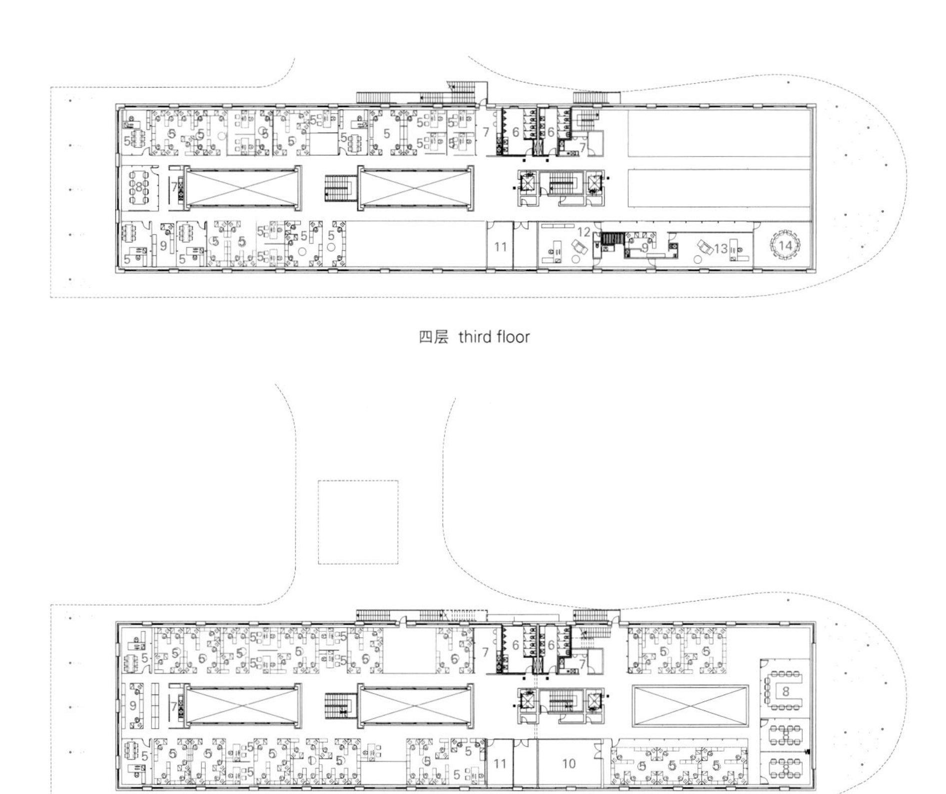

四层 third floor

三层 second floor

1. 停车场
2. 入口
3. 大厅
4. 会议室
5. 办公室
6. 休息室
7. 食品储藏室
8. 会议室
9. 秘书室
10. 计算机房
11. 储藏室
12. 名誉主席办公室
13. 主席办公室
14. 接待室
15. 副主席办公室

1. parking lot
2. entrance
3. lobby
4. meeting place
5. office
6. restroom
7. pantry
8. meeting room
9. secretary's office
10. computer room
11. storage
12. honorary president's room
13. president's room
14. reception room
15. secondary's room

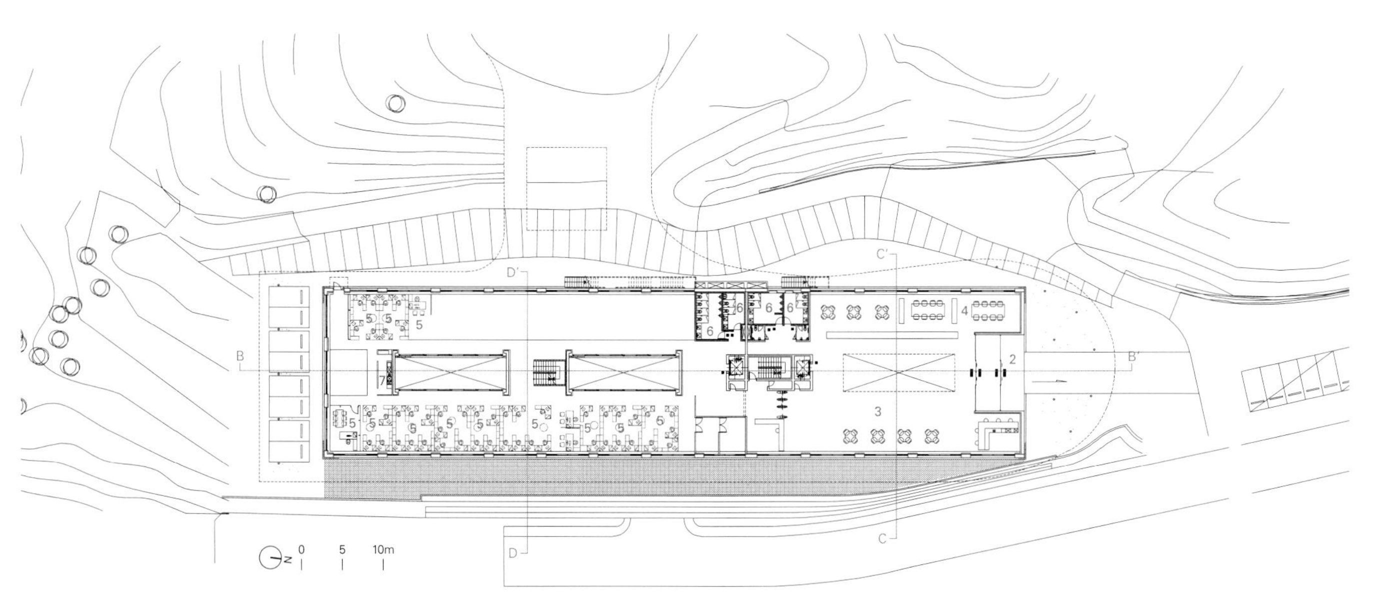

二层 first floor

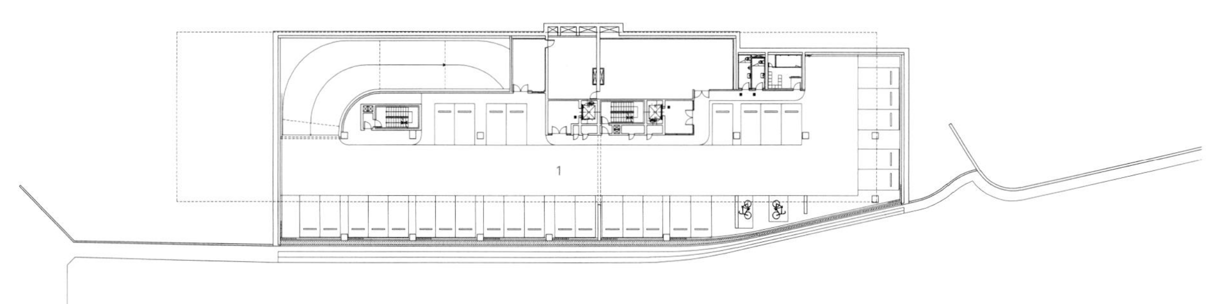

地下一层 first floor below ground

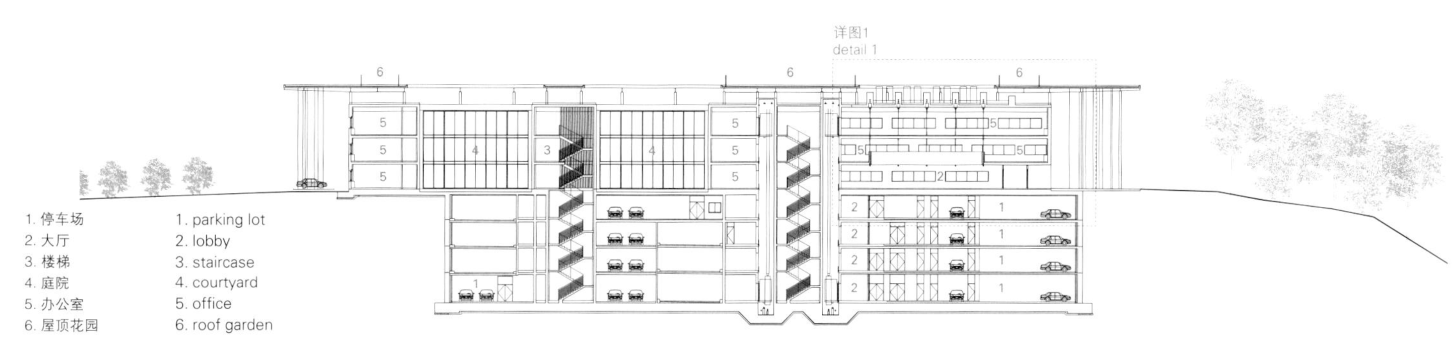

B-B' 剖面图 section B-B'

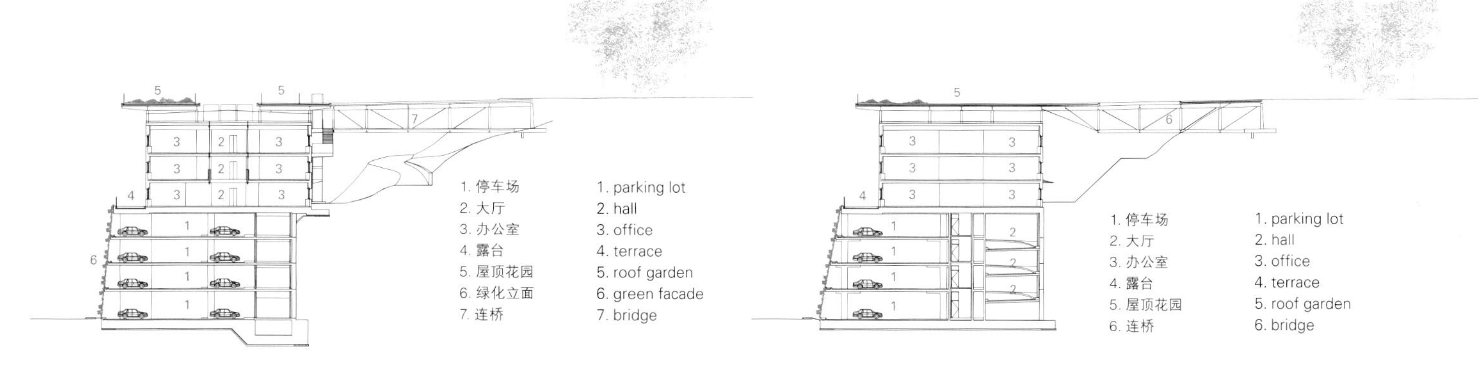

C-C' 剖面图 section C-C'

D-D' 剖面图 section D-D'

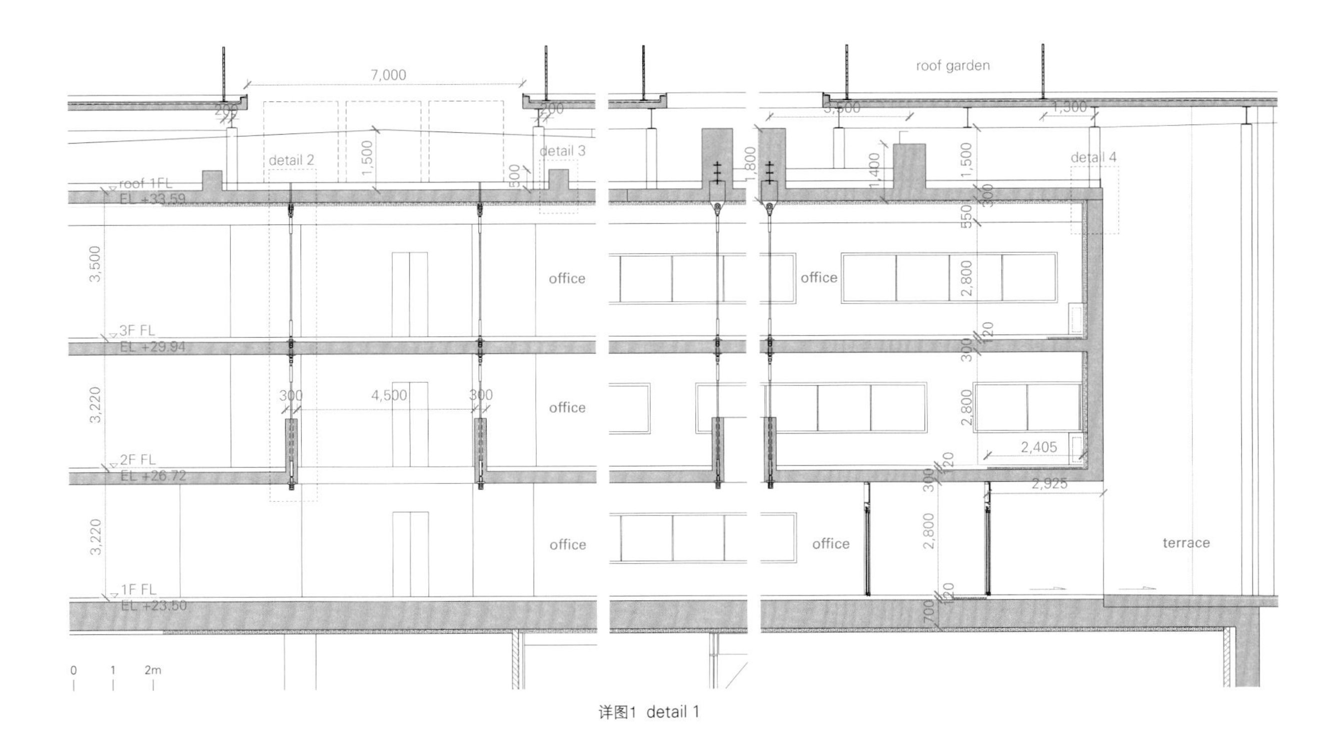
roof garden
7,000
detail 2
detail 3
detail 4
roof 1FL
EL +33.59
3F FL
EL +29.94
2F FL
EL +26.72
1F FL
EL +23.50
office
office
office
office
office
office
terrace
4,500
2,405
2,925
0 1 2m

详图1 detail 1

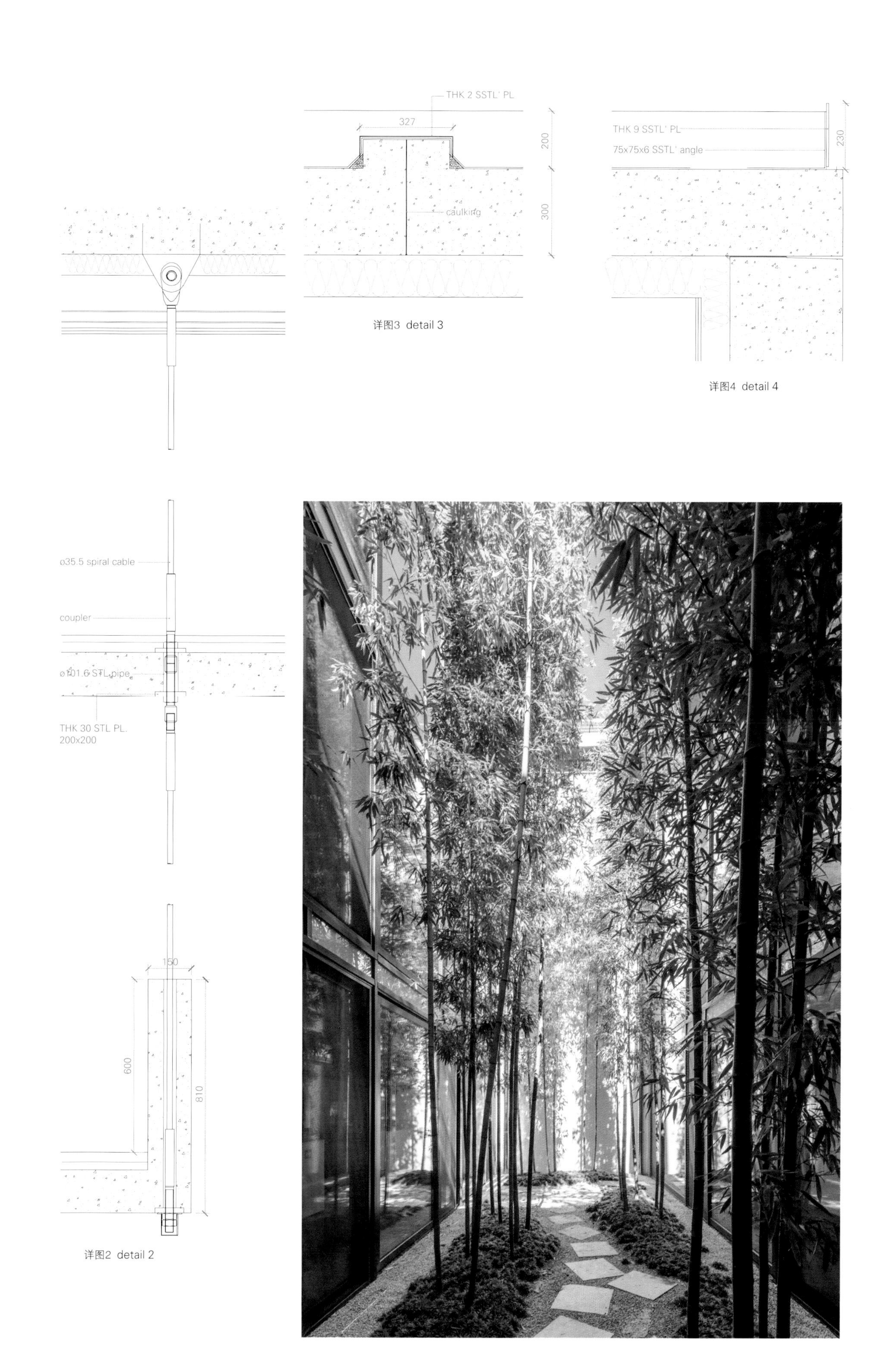

详图3 detail 3

详图4 detail 4

详图2 detail 2

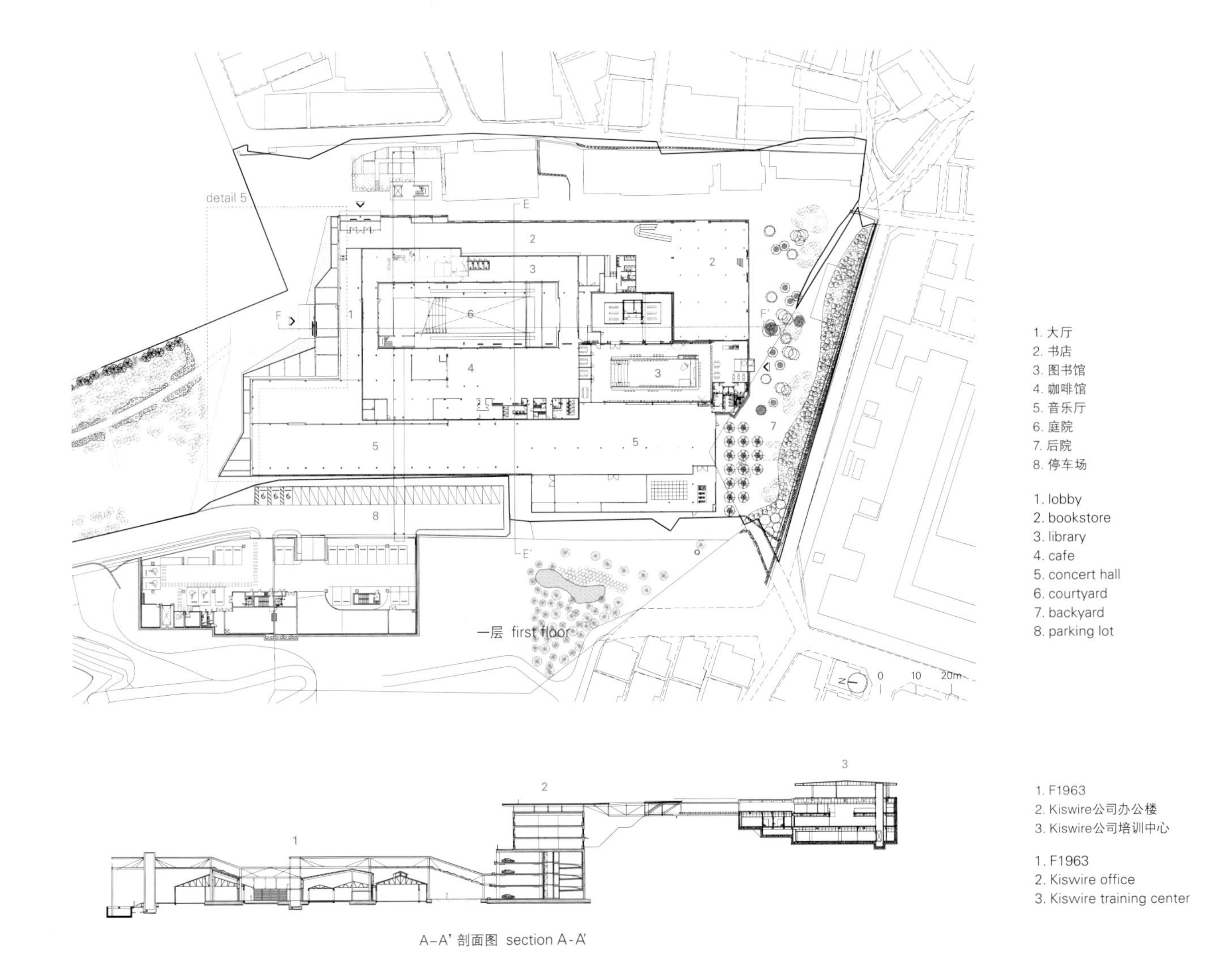

A–A' 剖面图 section A-A'

F1963

新循环建筑

F1963是位于釜山市Mangmidong小区Kiswire建筑群内的一个项目，包括对该公司在现场建造的、未使用的电线工厂进行改造。随着工厂的不断扩建，这座建筑在其悠久的历史中积累了新的痕迹。这个场地在这些不同历史时刻的改造中，讲述了一系列不同的故事。因此，该项目试图在这些叙述之间建立一种和谐的关系，用场地自身的历史来吸引对话。我们可以称之为"新循环建筑"，对旧事物进行创造性的重新诠释，用独特的记忆和时间痕迹来代替严格的保护或有怀旧性质的重建。F1963通过三种不同的建筑手法引入了这一概念：保存（利用现存的建筑特征）、减法（创建庭院空间和立面）和添加（使建筑的形式向外突出，创建一个覆盖着天蓝色网眼金属板的间隙空间）。

保存

在这个荒无人迹的工厂里，在其随意扩张的建筑历史中，我们发现了长期存在于油污、剥落的油漆和饱经风霜的木桁架上的历史痕迹，或者也许在工厂中心的一个发电机里，这个发电机很久以前就不再履行它的工业使命了。F1963试图以人们可以接受的、最低限度的建筑干预方式来激活这段历史。在这种气氛之下，无论是通过庭院体量的切割而暴露的木桁架，还是沿着地面重新展示的混凝土，这些贯穿整座建筑的各种对象，都找到了新的生命。在入口处，这些元素成为家具和实际互动的对象，划分了发现和人为设计之间的边界。

减法

商业体量是该方案唯一的建筑附加元素，它贯穿室内庭院，希望通过一个非正式的展览空间，鼓励游客体验现有工厂的空间和时间层次。这种通过原始结构中心进行切割的操作创造了一个空间，确保日光可以穿透进入室内，也确保了场地的被动通风。在建筑的正面，一面天蓝色的网眼金属板墙模仿了原始结构的形状，仿佛在建筑中为充满可能性和丰富体验的空间的最终扩建带来了一种富于想象力的诠释。

添加

F1963的入口是一个由天蓝色的网眼金属板包裹的突出体量。这种形式设计手法中和了现有结构的形象质量，然而，色彩的介入使得访客对新旧两个区域的真实性产生了质疑。它创造了一个空间，在这个空间里，一天中逐渐变化的光线会反射到灵活空间的金属板上，并在柔和的光线中变得更加和谐，就像旁边的竹林被微风吹得沙沙作响。透过金属板上的金属槽为室内空间遮挡光线，这种设计营造了一个吸引游客的奇妙空间。

同样，横跨建筑设施的连桥也被涂成了蓝色，这是为了参照方案中附加的建筑特征。它不仅仅是一个从办公室到文化空间的通道，也是可以欣赏周围地区景色的地方，这样一来，也为整个Kiswire建筑群增添了一些完全独特的氛围。

F1963

Newcycle Architecture

A project within the Kiswire compound in the Mangmi-dong neighborhood of Busan, F1963 comprises a renovation of the company's unused wire factory built on site. As the factory was expanded throughout its life, the building accumulated new traces of its long history. Palimpsestic in its layering of these various historical moments, the site tells a diverse series of stores. The project therefore attempts to arbitrate a kind of harmony between these narratives, enticing the conversation the site poses with its own history. We might term this "Newcycle Architect", a creative reinterpretation of old things which hold unique traces of memory and time in place of strict preservation or nostalgic reconstruction. F1963 engages this concept through three distinct architectural maneuvers: conservation (using the extant architectural features), subtraction (creating the courtyard space and facade) and addition (projecting the form of the facility outwards to create an interstitial space clad in sky-blue expanded metal).

Conservation

In this desolate factory, layered in its history of haphazard architectural expansions, we find testimonies of time long endured in oil stains, stripping paint and weathered wooden trusses, or perhaps in a generator at the heart of the facility which long ago relinquished its industrial responsibilities. F1963 endeavors to provoke these histories with the most acceptably minimal architectural interven-

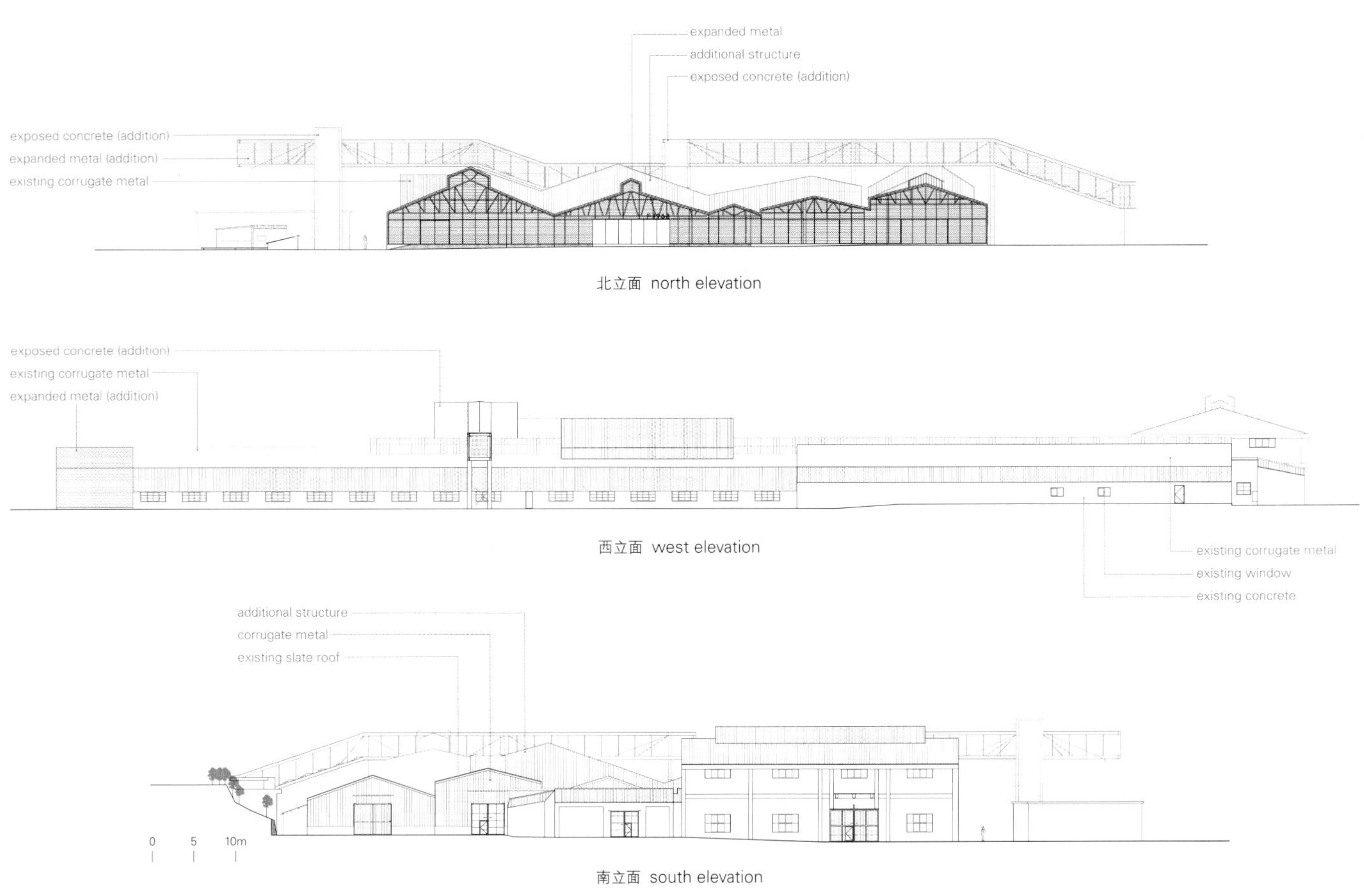

北立面 north elevation

西立面 west elevation

南立面 south elevation

项目名称：Kiswire F1963 / 地点：20, Gurak-ro 123beon-gil, Suyeong-gu, Busan, Korea / 建筑师：BCHO Architects / 项目团队&施工监理：Kwon Do-yeon, Choi Ha-young, Shin Myung, Cha Yoon-ji / 施工：Hakrim construction & engineering / 客户：Kiswire / 用途：cultural complex and commercial / 用地面积：21,963.90m² / 建筑面积：10,184.51m² / 总建筑面积：11,005.83m² / 结构：steel frame / 设计时间：2015.12—2016.2
施工时间：2016.3—2016.8 / 摄影师：©Sergio Pirrone (courtesy of the architect)

Hybridizing Earth Discussing Multitude
2016 부산비엔날레
Busan Biennale 2016
9/3—11/30
Busan Museum of Art X F1963
www.busanbiennale.org
혼혈하는 지구
다중지성의 공론장
F

부산비엔날레
Biennale 2016

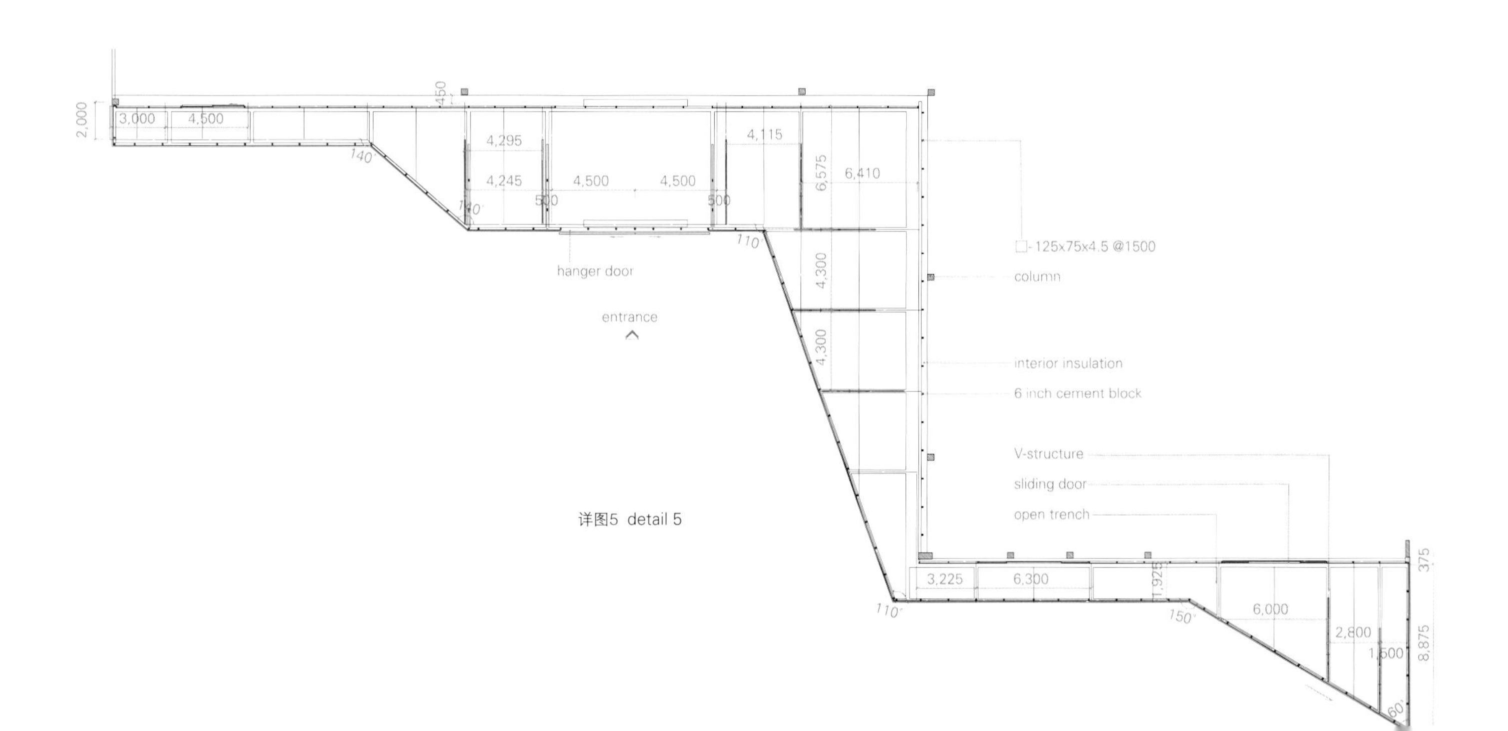
2,000
450
3,000
4,500
140
4,295
4,245
4,500
4,500
500
500
140
4,115
6,575
6,410
110
hanger door
entrance
4,300
4,300
□-125x75x4.5 @1500
column
interior insulation
6 inch cement block
V-structure
sliding door
open trench
3,225
6,300
1,925
110°
150°
6,000
2,800
1,600
375
8,875
60°

详图5 detail 5

tion. In this tone, wooden trusses exposed by the slicing of the courtyard volume or concrete reshown along the floor find new life as various objects throughout the building. At the entrance, such elements become furniture, physically interactive objects which demarcate the boundary between that which is found and that which is contrived.

Subtraction

The commercial volume, the only architecturally additive element of the scheme, penetrates the interior courtyard in the hopes of encouraging visitors to experience the spatial and temporal hierarchy of the existing factory by walking around an informally positioned exhibition space. This operation of slicing through the center of the original

structure crafts a space which secures daylight penetration and passive ventilation across the site. At the front, a light blue wall of expanded metal mimics the figure of the original structure as if to create in the building itself an imagined interpretation of its eventual expansion forwards a space of possibility and experience.

Addition

The entrance of F1963 is a projected volume sheathed in bright blue expanded metal. This formal maneuver compromises between the figural qualities of the existing structure and yet brings a chromatic intervention which makes the visitor question the authenticity of both areas: of the new and of the old. It crafts a space in which the gradual change of light across a day reflects luminously off the metal in the flexible space and harmonizes in its soothing light as does an adjacent bamboo grove to the gentle rustle of a passing breeze. The dim screening of its interior space through the metallic slots in its panels creates a curious space which draws visitors in.

Likewise, the bridge spanning the facility is indexical painted blue to reference its additive architectural identity in the scheme. It aspires to be more than a mere passage from the offices to the cultural spaces but as a place from which to take in the views of surrounding areas, in so doing adding something wholly unique to the atmosphere of the entire Kiswire compound.

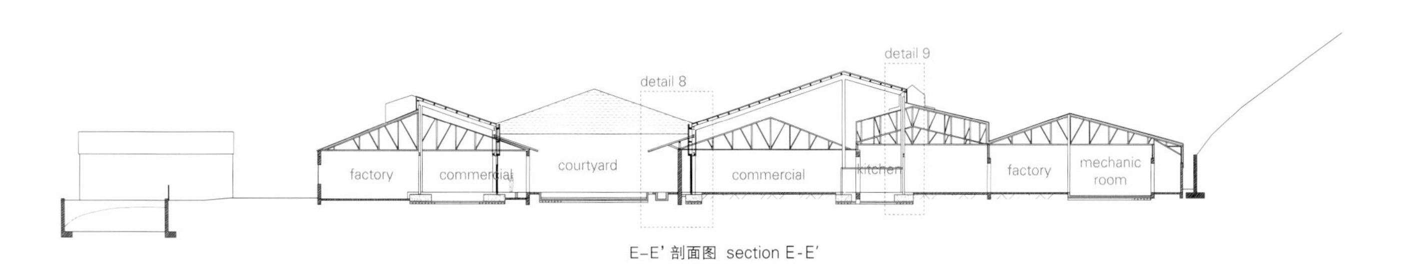

E-E' 剖面图 section E-E'

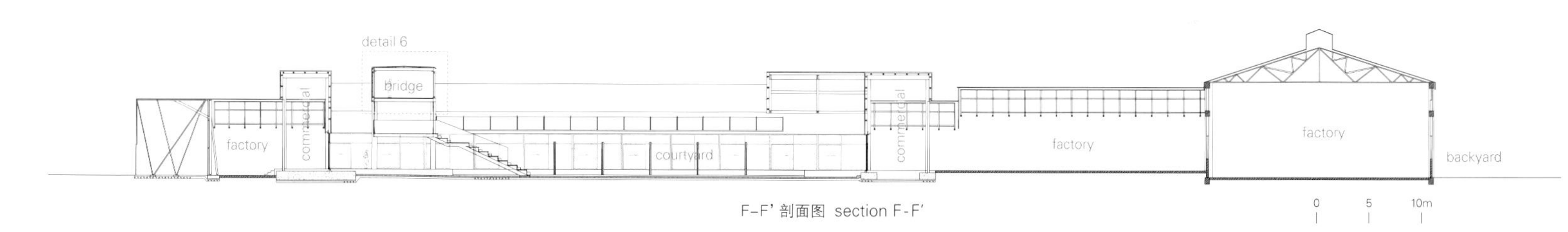

F-F' 剖面图 section F-F'

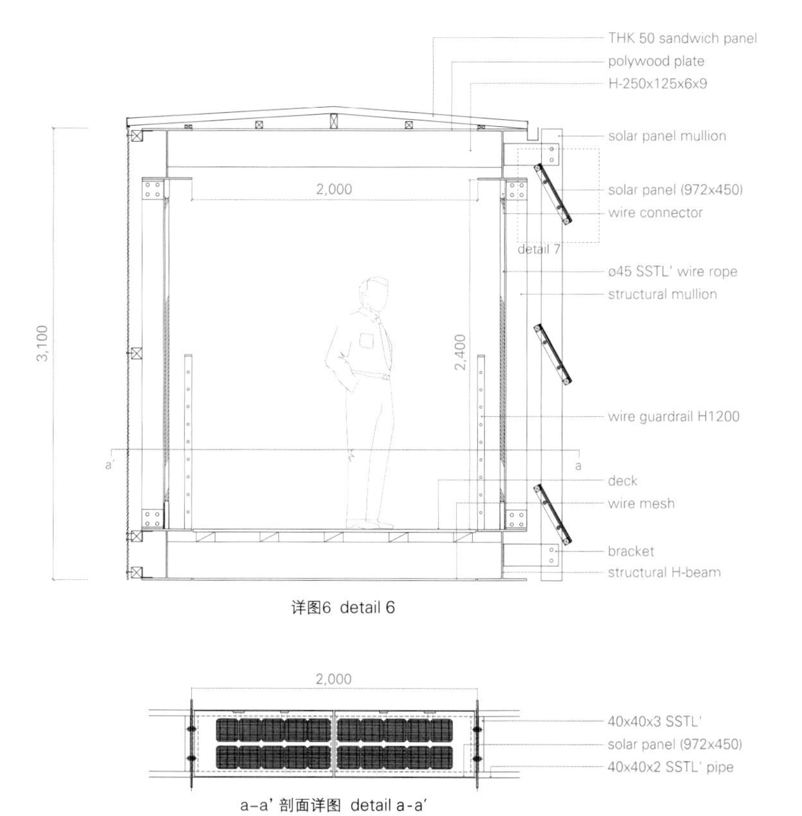

详图6 detail 6

2,000

40x40x3 SSTL'
solar panel (972x450)
40x40x2 SSTL' pipe

a–a' 剖面详图 detail a-a'

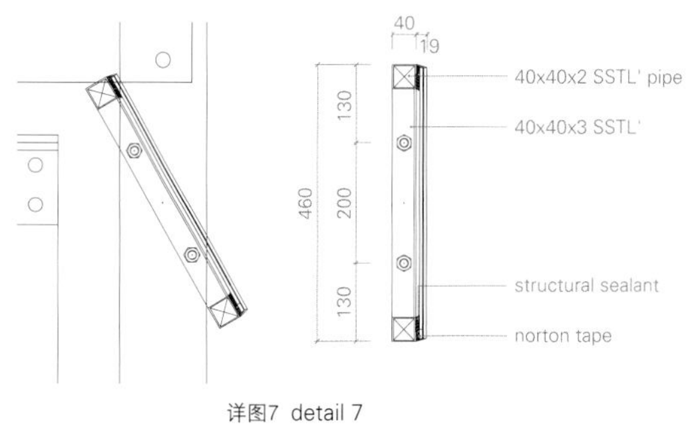

详图7 detail 7

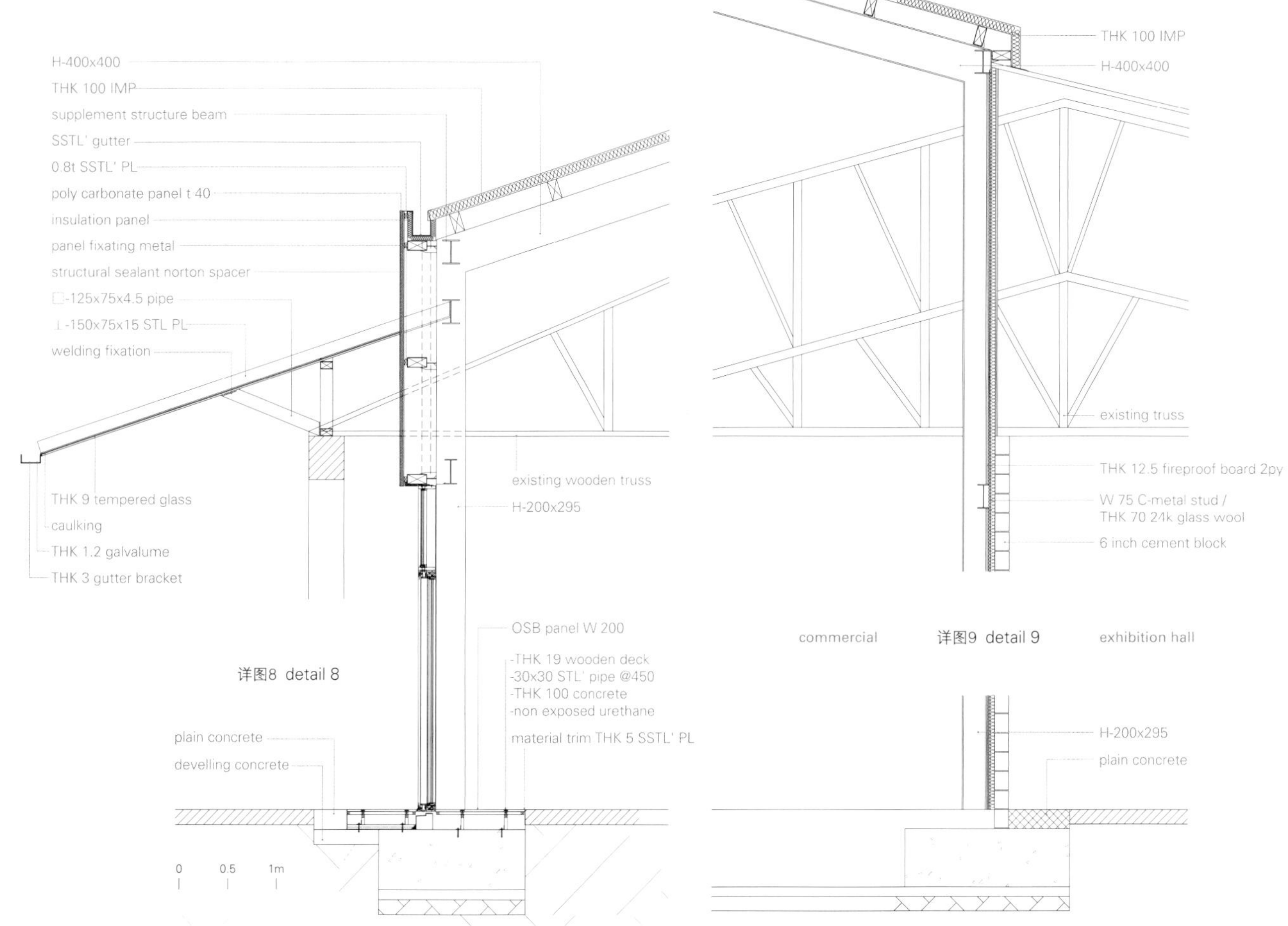

详图8 detail 8

详图9 detail 9

乐高之家

LEGO House

BIG

“乐高之家是乐高积木具有无限可能性的真实体现。通过其系统的创造力，所有年龄段的孩子都拥有了创造自己世界的能力，并在游戏的时候‘居住’在他们自己的世界里。这就是建筑和乐高游戏的精髓所在：使人们能够想象比现状更令人兴奋、更富表现力的新世界，并为他们提供使之成为现实的技能。这就是孩子们每天用乐高积木做的事情——这也是我们今天在乐高之家用真正的砖块做的事情，让丹麦的比伦德离成为儿童之都又近了一步。”BIG的创始合伙人比雅克·英格尔斯表示。

BIG和乐高公司将经典乐高积木的玩具比例代入乐高之家的建筑比例，形成了巨大的展览空间和公共广场，体现了乐高所有体验的核心文化和价值观。由于23m高的乐高之家坐落在比伦德的中央，它被设想为一个城市空间和体验中心。21个重叠的块体像独立的建筑一样放置在那里，形成了一个2000m²的乐高广场，通过体块之间的缝隙得到自然采光。该广场看起来像一个没有任何可见柱子的城市洞穴，公众可以随意进入，譬如，比伦德的游客和市民可以横穿建筑，把这里当成捷径。乐高广场充满了城市活力，欢迎当地人和游客来到咖啡馆、餐厅、乐高商店和会议设施。在广场上方，一组画廊重叠在一起，创造出了一连串的展览空间。每个画廊都用乐高积木的原色进行了颜色编码，所以在展览中寻找方向就变成了一段穿越色彩光谱的旅程。

一层和二层包括四个按颜色排列的游戏区，并按照代表孩子们某一学习方面的活动进行功能设计：红色代表创造性，蓝色代表认知，绿色代表社交，黄色代表情感。所有年龄层的客人都可以拥有身临其境的互动体验，表达他们的想象力，尤其是遇到来自世界各地的其他建设者的挑战。该建筑的顶部是杰作画廊，这是乐高爱好者喜爱的作品集合，也是在向乐高社区致敬。杰作画廊由标志性的2×4乐高积木组成，在8个类似积木钉的圆形天窗下展示艺术。就像黄金比例一样，积木的比例被嵌入建筑中所有人造物体的几何形状中，从台阶和墙壁上的釉面瓷砖到21个体块的整体设计方案。在这座杰作画廊的顶层，市民和游客可以360度全景欣赏这座城市。屋顶的一部分可以通过像素化的公共楼梯进入，楼梯可以兼作非正式的礼堂，供观赏风景或观看

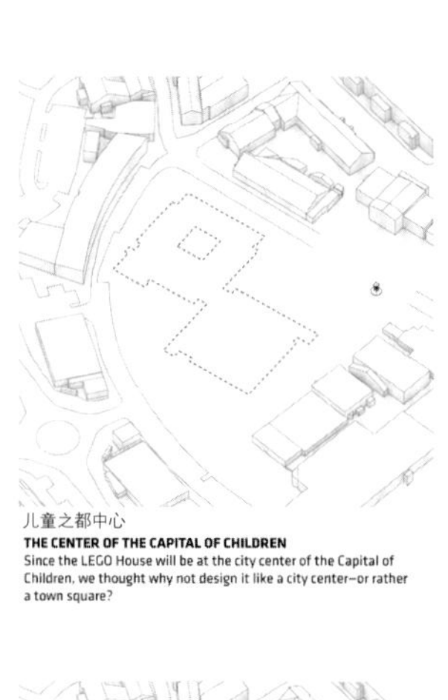

儿童之都中心
THE CENTER OF THE CAPITAL OF CHILDREN
Since the LEGO House will be at the city center of the Capital of Children, we thought why not design it like a city center—or rather a town square?

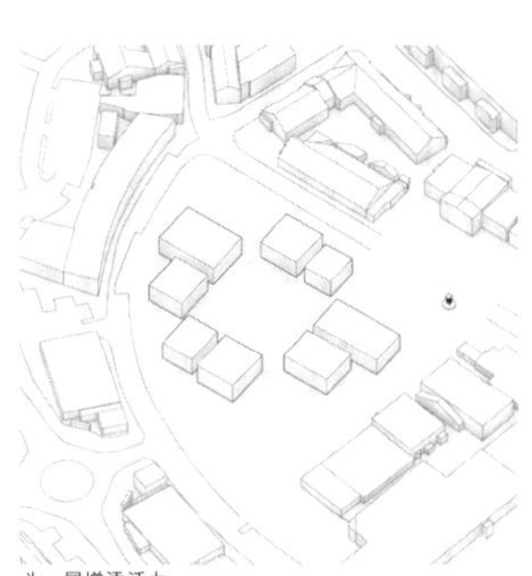

为一层增添活力
ENERGIZING THE GROUND LEVEL
We consolidated all the elements of the program that have an outward-oriented, everyday-like urban character around a central space: café, forum, LEGO store, ticket offices, wardrobe and restrooms, offices and loading.

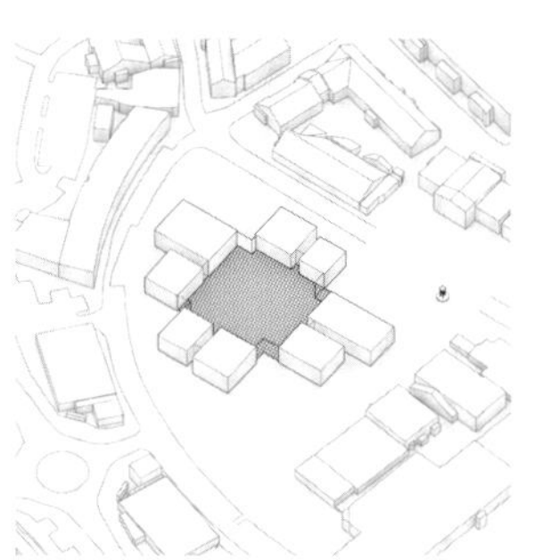

儿童广场
CHILDREN'S SQUARE
Placed like individual buildings framing a square, they allow daylight and views to pass between them while letting people enter from multiple directions and allowing shortcuts through the building—like crossing a plaza.

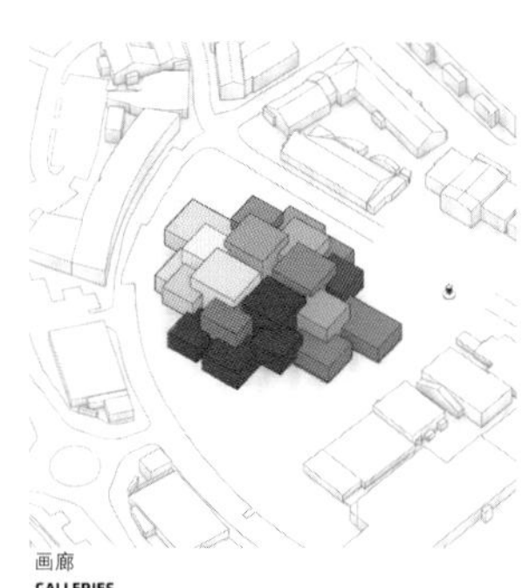

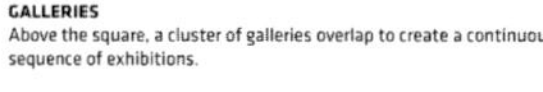

画廊
GALLERIES
Above the square, a cluster of galleries overlap to create a continuous sequence of exhibitions.

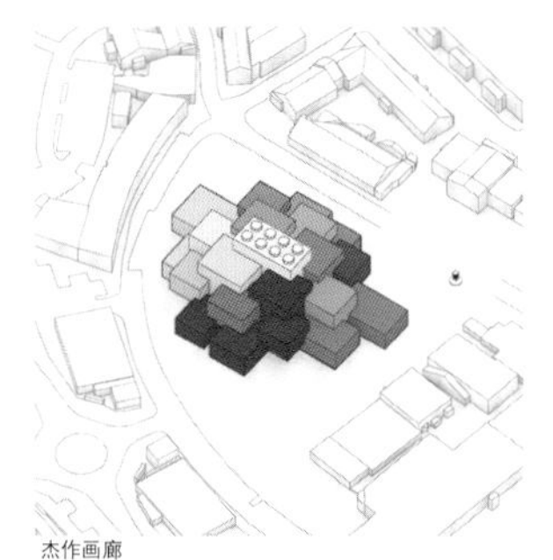

杰作画廊
MASTERPIECE GALLERY
At the top of the pile of bricks the Masterpiece Gallery provides a bridge between all the corners of the exhibition, and serves as a sky-lit gallery for LEGO as an art form.

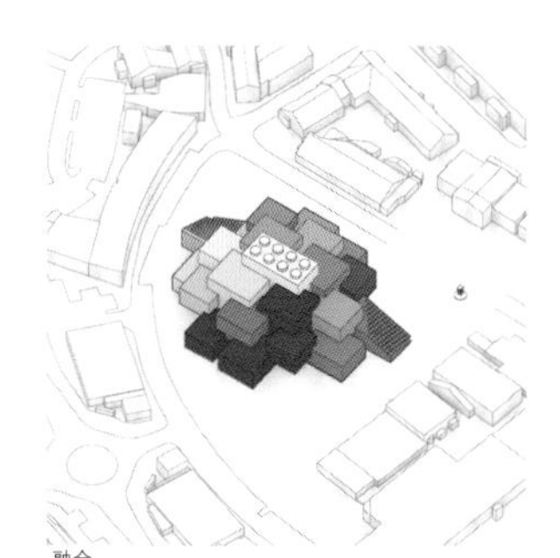

融合
MELT
Two of the volumes seem to melt in a pixelated way to form informal auditoria for people watching or public performances.

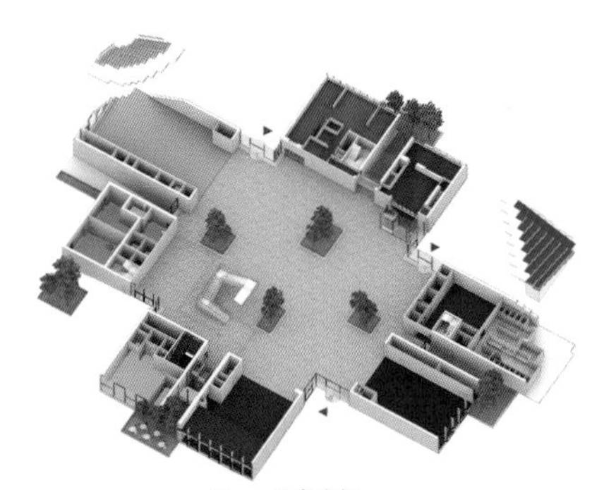

一层——儿童广场
ground level-children's square

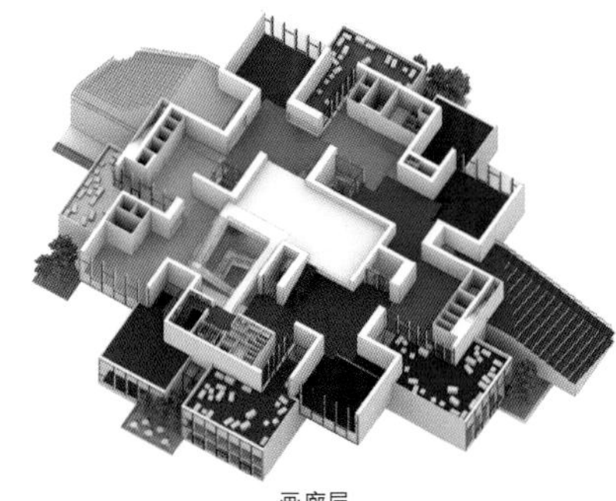

画廊层
galleries level

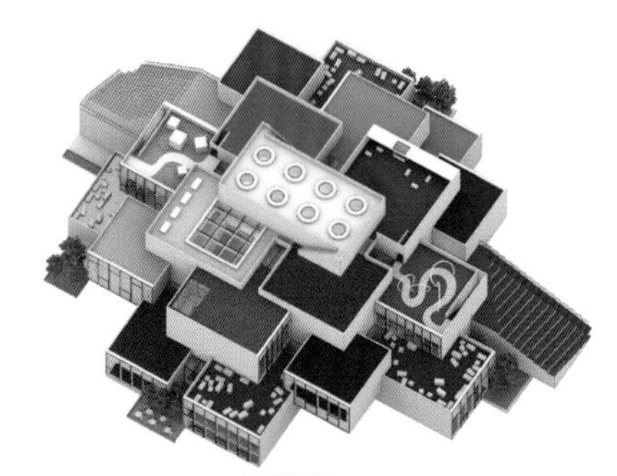

屋顶景观
roofscape

演出的人坐下。在位于较低楼层的历史收藏馆里，游客可以体验到乐高公司的历史及其品牌故事。这个位于乐高广场下面的拱顶，是孩子们和乐高爱好者（成年乐高迷）见证乐高历史上几乎每一套乐高玩具第一个版本的地方，包括模仿乐高之家积木结构的774件、197步的新玩具套装。

"LEGO House is a literal manifestation of the infinite possibilities of the LEGO brick. Through systematic creativity children of all ages are empowered with the tools to create their own worlds and to inhabit them through play. At its finest – that is what architecture – and LEGO play – is all about: enabling people to imagine new worlds that are more exciting and expressive than the status quo – and to provide them with the skills to make them reality. This is what children do every day with LEGO bricks – and this is what we have done today at LEGO House with actual bricks, taking Billund a step closer towards becoming the Capital for Children." says Bjarke Ingels, Founding Partner, BIG.

BIG and LEGO bring the toy scale of the classic LEGO brick to architectural scale with LEGO House, forming vast exhibition spaces and public squares that embody the culture and values at the heart of all LEGO experiences. Due to its central location in the heart of Billund, the 23m tall LEGO House is conceived as an urban space as much as an experience center. 21 overlapping blocks are placed like

©Kim Christensen(courtesy of the architect)

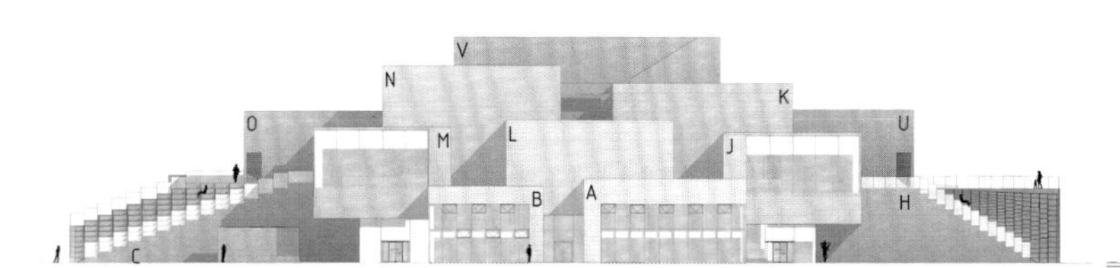

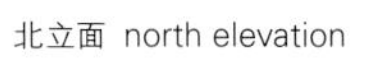

北立面 north elevation

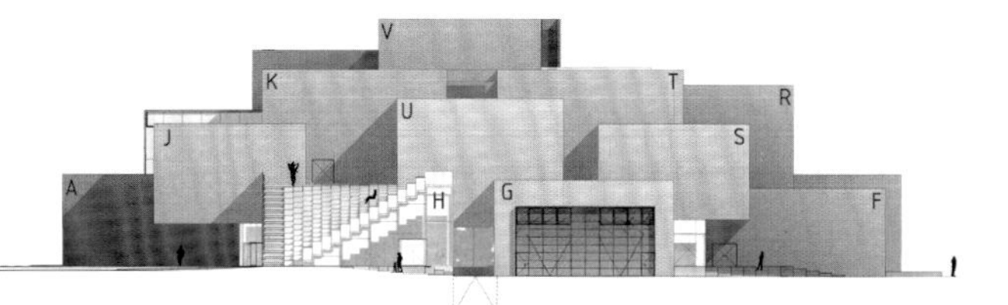

西立面 west elevation

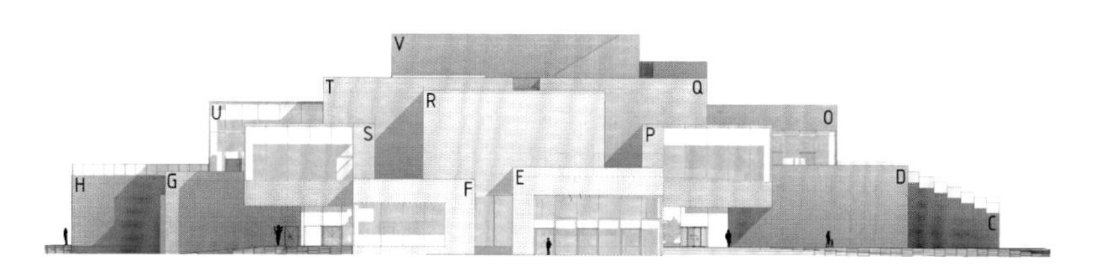

南立面 south elevation

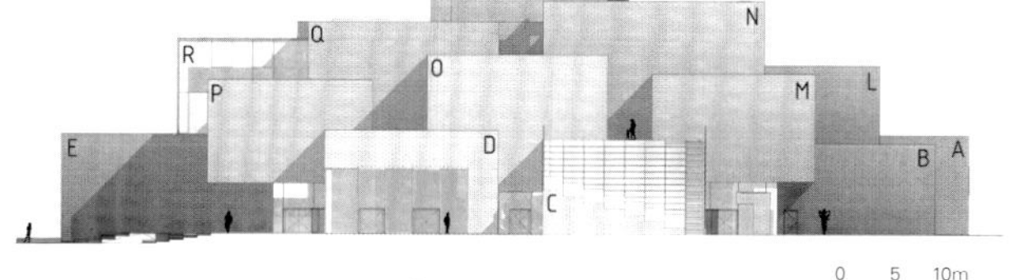

东立面 east elevation

COFFEE
JUICE
SMOOTHIES
PANINI
AMERICANO 30,-
ESPRESSO 25,-
DB.ESPRESSO 30,-
CORTADO 38,-
CAFE LATTE 45,-
CAPPUCCINO 45,-
CHOCOLATE 45,-
OVERPRICED COFFEE
42,-
42,-
59,-
49,-
BRICKACCINO

项目名称：LEGO House / 地点：Billund, Denmark
建筑师：BIG
主管合伙人：Bjarke Ingels, Finn Nørkjær, Brian Yang
项目主管：Brian Yang / 项目经理：Finn Nørkjær
项目建筑师&立面：Snorre Nash
项目团队：Andreas Klok Pedersen, Agne Tamasauskaite, Annette Birthe Jensen, Ariel Joy Norback Wallner, Ask Hvas, Birgitte Villadsen, Chris Falla, Christoffer Gotfredsen, Daruisz Duong Vu Hong, David Zahle, Esben Christoffersen, Franck Fdida, Ioana Fartadi Scurtu, Jakob Andreassen, Jakob Ohm Laursen, Jakob Sand, Jakub Matheus Wlodarczyk, Jesper Bo Jensen, Jesper Boye Andersen, Julia Boromissza, Kasper Reimer Hansen, Katarzyna Krystyna Siedlecka, Katarzyna Stachura, Kekoa Charlot, Leszek Czaja, Lone Fenger Albrechtsen, Louise Bøgeskov Hou, Mads Enggaard Stidsen, Magnus Algreen Suhr, Manon Otto, Marta Christensen, Mathias Bank Stigsen, Michael Kepke, Ole Dau Mortensen, Ryohei Koike, Sergiu Calacean, Søren Askehave, Stefan Plugaru, Stefan Wolf, Thomas Jakobsen Randbøll, Tobias Hjortdal, Tommy Bjørnstrup
合作方：COWI, Dr. Lüchinger+Meyer Bauingenieure, Jesper Kongshaug, Gade & Mortensen Akustik, E-Types
客户：LEGO
总建筑面积：12,000m² / 竣工时间：2017
摄影师：©Iwan Baan (courtesy of the architect) (except as noted)

individual buildings, framing a 2,000m² LEGO square that is illuminated through the cracks and gaps between the volumes. The plaza appears like an urban cave without any visible columns and is publicly accessible, allowing visitors and citizens of Billund to shortcut through the building. The LEGO square is energized by an urban character, welcoming locals and visitors to the café, restaurant, LEGO store and conference facilities. Above the square, a cluster of galleries overlap to create a continuous sequence of exhibitions. Each gallery is color-coded in LEGO's primary colors so wayfinding through the exhibitions becomes a journey through the color spectrum.
The first and second floors include four play zones arranged by color and programmed with activities that represent a

certain aspect of a child's learning: red is creative, blue is cognitive, green is social, and yellow is emotional. Guests of all ages can have an immersive and interactive experience, express their imagination, and not least be challenged by meeting other builders from all over the world. The top of the building is crowned by the Masterpiece Gallery, a collection of LEGO fans' beloved creations that pay tribute to the LEGO community. The Masterpiece Gallery is made of the iconic 2x4 LEGO brick and showcases art beneath eight circular skylights that resemble the studs of the brick. Like the golden ratio, the proportions of the brick are nested in the geometries of everything man-made in the building, from the glazed ceramic tiles in the steps and walls to the overall 21 block scheme. Atop the Masterpiece Gallery, citizens and visitors can get a 360° panoramic view of the city. Some of the rooftops can be accessed via pixelated public staircases that double as informal auditoria for people watching or seating for performances. The History Collection at the lower level is where visitors can experience an archival immersion into the LEGO company and brand's story. The Vault – located underneath LEGO Square – is where children and AFoLs (Adult Fans of LEGO) can witness the first edition of almost every LEGO set ever manufactured, including the new 774-piece, 197-step kit replicating the stacked-block formation of the LEGO House.

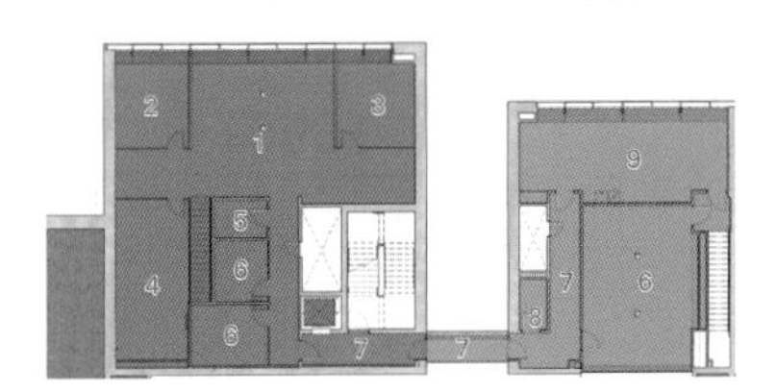

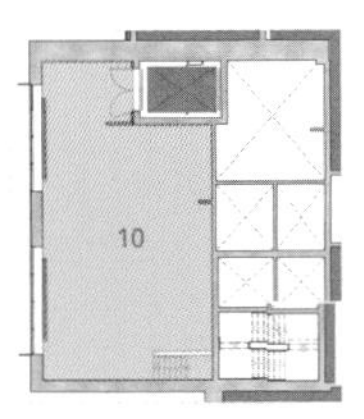

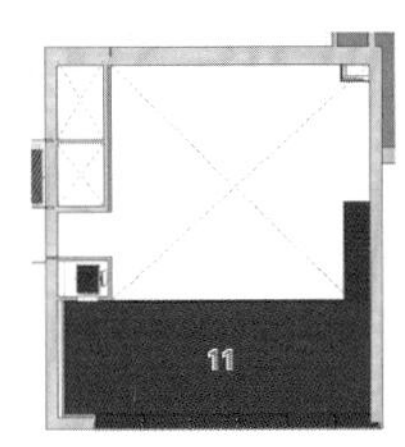

1. 行政管理区
2. 首席运营官办公室
3. 首席执行官办公室
4. 午餐室
5. 印刷/库房
6. 会议室
7. 通道
8. 卫生间
9. 人事办公室
10. 工作坊
11. 餐厅

1. administration
2. COO
3. CEO
4. lunch room
5. print/depot
6. meeting room
7. passage
8. toilet
9. personnel room
10. workshop
11. restaurant

夹层
mezzanine floor

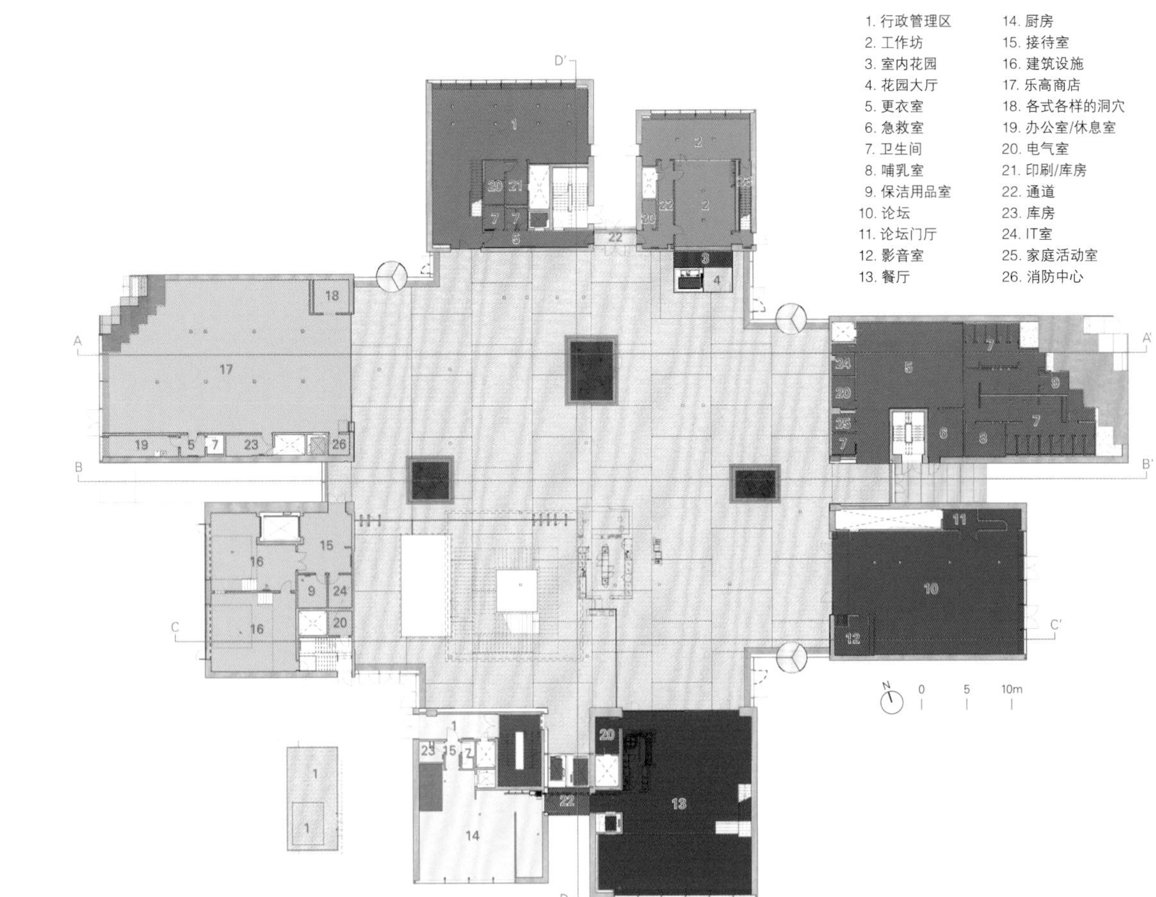

1. 行政管理区
2. 工作坊
3. 室内花园
4. 花园大厅
5. 更衣室
6. 急救室
7. 卫生间
8. 哺乳室
9. 保洁用品室
10. 论坛
11. 论坛门厅
12. 影音室
13. 餐厅
14. 厨房
15. 接待室
16. 建筑设施
17. 乐高商店
18. 各式各样的洞穴
19. 办公室/休息室
20. 电气室
21. 印刷/库房
22. 通道
23. 库房
24. IT室
25. 家庭活动室
26. 消防中心

1. administration
2. workshop
3. interior garden
4. garden foyer
5. wardrobe
6. first aid room
7. toilet
8. nursing room
9. janitorial room
10. forum
11. forum lobby
12. AV room
13. restaurant
14. kitchen
15. anteroom
16. building facilities
17. LEGO store
18. cave/diverse
19. office/break
20. elec. room
21. print/depot
22. passage
23. depot
24. IT-room
25. family room
26. fire central

一层
ground floor

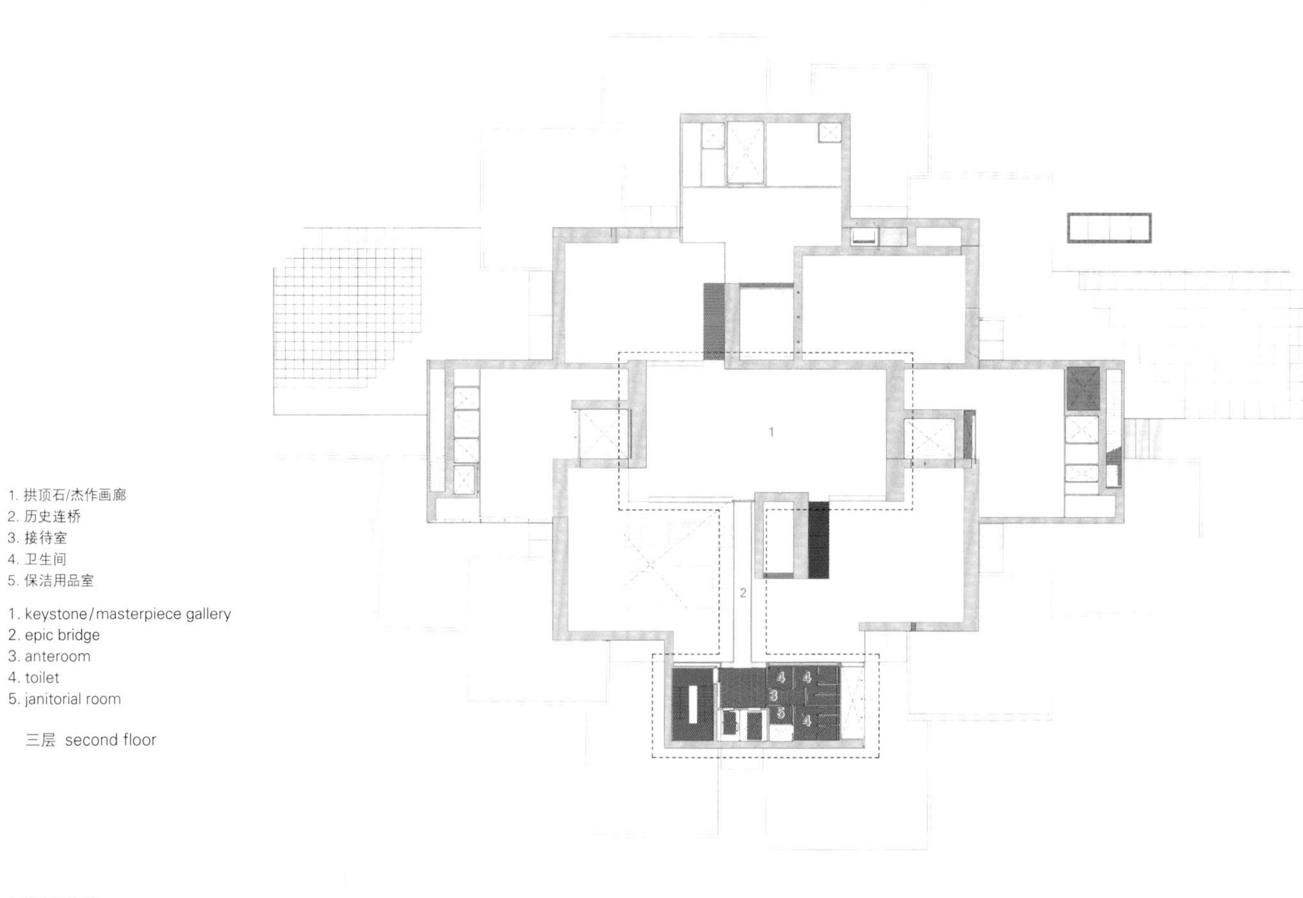

1. 拱顶石/杰作画廊
2. 历史连桥
3. 接待室
4. 卫生间
5. 保洁用品室

1. keystone/masterpiece gallery
2. epic bridge
3. anteroom
4. toilet
5. janitorial room

三层 second floor

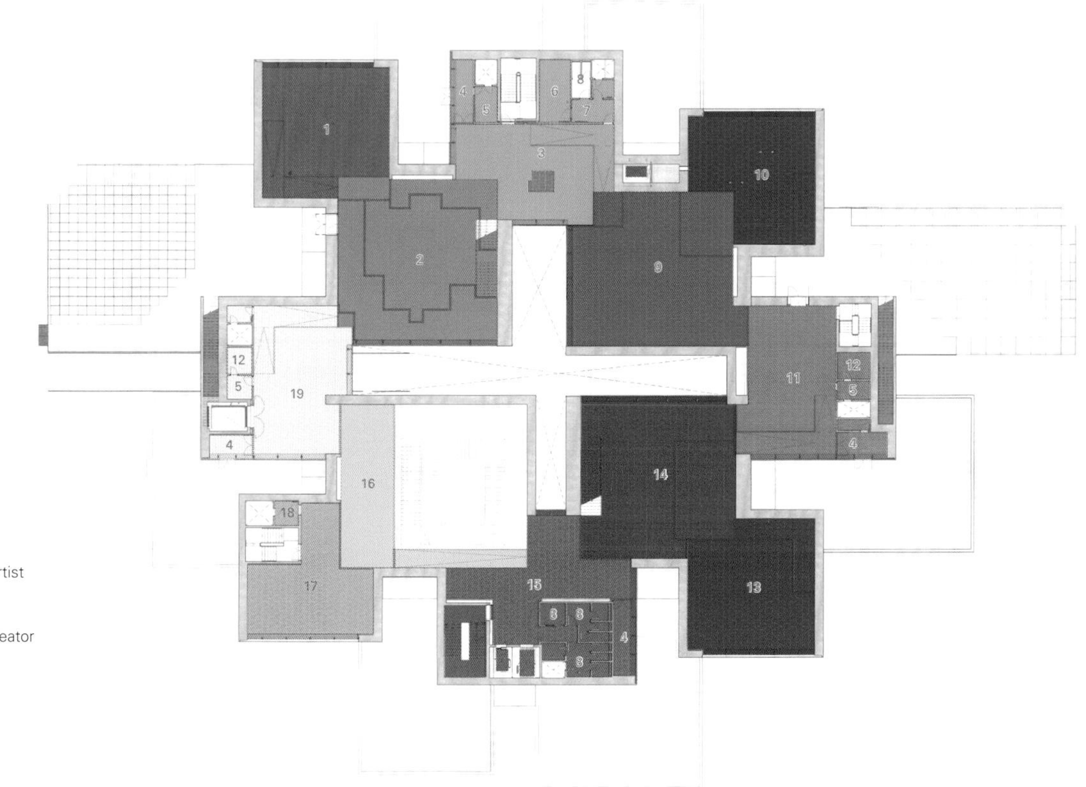

1. 故事实验室
2. 世界探索家
3. 玩具人偶创造者
4. 入口门厅
5. 电气室
6. 保洁用品室
7. 接待室
8. 卫生间
9. 城市建筑师
10. Robo实验室
11. 试车驾驶员
12. IT室
13. 创意实验室
14. UKTRYK积木建造者
15. UKTRYK门厅
16. 设计与制作/花朵艺术家
17. 小鱼设计师
18. 车库
19. 设计与制作/生物创造者

1. story lab
2. world explorer
3. minifigure creator
4. entrance foyer
5. electronic room
6. janitorial room
7. anteroom
8. toilet
9. city architect
10. Robo lab
11. test driver
12. IT-room
13. creative lab
14. UKTRYK/brick builder
15. UKTRYK/lobby
16. design & fabrication/flower artist
17. fish designer
18. garbage
19. design & fabrication/critter creator

二层 first floor

Toilets
LEGO
FASCINATING
History Collection

1. 乐高商店 2. 储藏室 3. 停车场 4. 技术设备间 5. 世界探索者 6. 更衣室 7. 卫生间 8. 屋顶露台

1. LEGO store 2. storage 3. parking place 4. technical room 5. world explorer 6. wardrobe 7. toilet 8. roof terrace

A-A' 剖面图 section A-A'

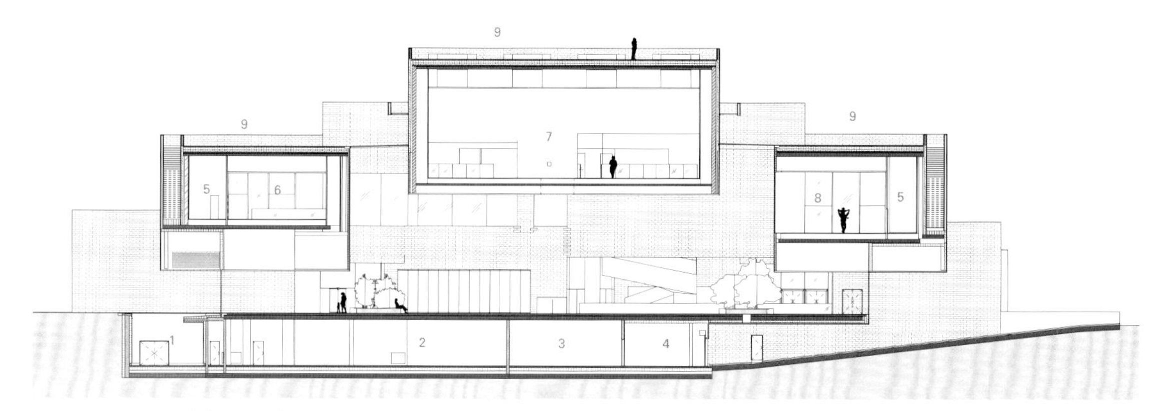

1. 发电机轴 2. 技术设备间 3. 工作坊 4. 停车场 5. 电气室 6. 试车驾驶员 7. 拱顶石/杰作画廊 8. 设计与制作/生物创造者 9. 屋顶露台

1. generator shaft 2. technical room 3. workshop 4. parking place 5. elec. room 6. test driver
7. keystone/masterpiece gallery 8. design&fabrication/critter creator 9. roof terrace

B-B' 剖面图 section B-B'

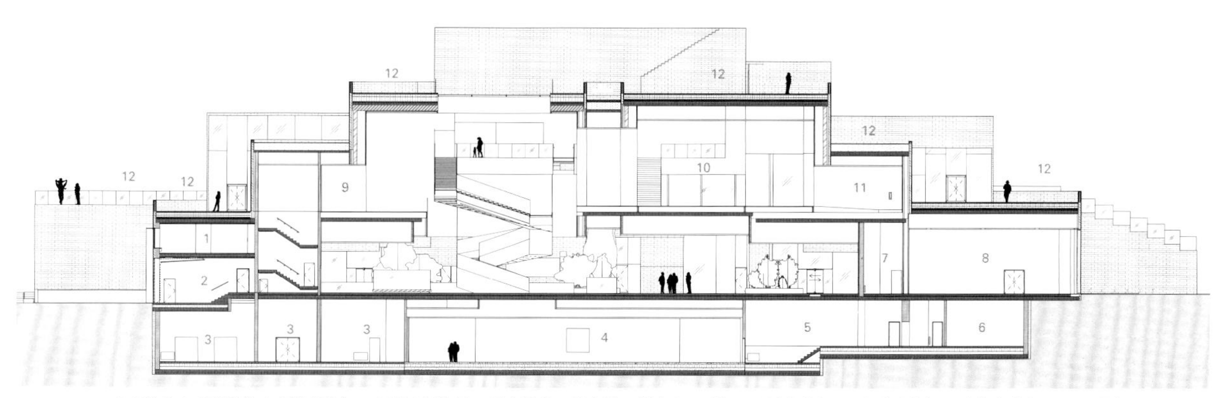

1. 工作坊 2. 建筑设施 3. 展览储藏室 4. 历史区/展览区 5. 逃生通道 6. 服务器 7. 影音室 8. 论坛 9. 小鱼设计师 10. 积木建造者 11. 创意实验室 12. 屋顶露台

1. workshop 2. building facilities 3. exhibition storage 4. history zone/exhibition 5. escape route passage
6. server 7. AV room 8. forum 9. fish designer 10. brick builder 11. creative lab 12. roof terrace

C-C' 剖面图 section C-C'

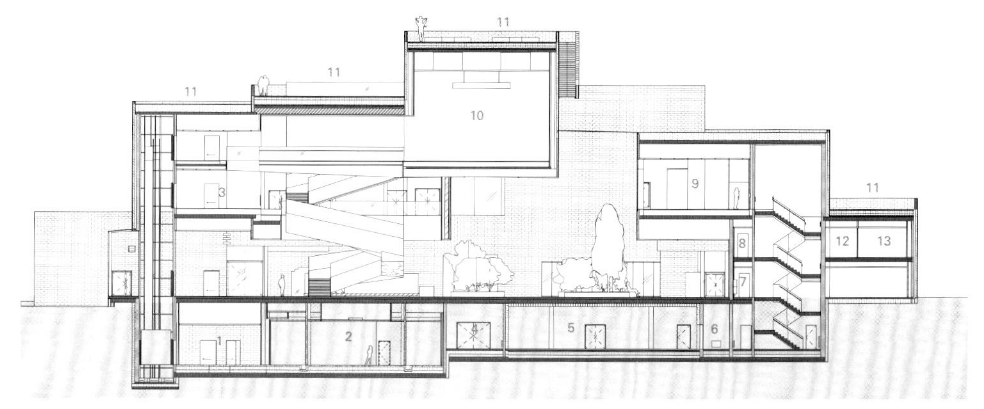

1. 门厅 2. 历史区/展览区 3. UKTRYK门厅 4. 工作坊 5. 停车场 6. 中央喷淋系统 7. 更衣室
8. 通道 9. 玩具人偶创造者 10. 拱顶石/杰作画廊 11. 屋顶露台 12. 行政夹层 13. 首席执行官办公室

1. foyer 2. history zone/exhibition 3. UDTRYK/lobby 4. workshop 5. parking place 6. sprinkle central 7. wardrobe
8. passage 9. minifigure creator 10. keystone/masterpiece gallery 11. roof terrace 12. administration mezzanine 13. CEO

D-D' 剖面图 section D-D'

Natura 公司圣保罗总部

Natura Headquarters São Paulo

Dal Pian Arquitetos

Natura公司成立于1969年，是巴西一家经营化妆品和美容产品的跨国公司，其产品在全球70个国家的3200多家门店销售。

2011年，Natura公司发起了一场建筑设计竞赛，目的是设计位于圣保罗的新行政总部。这个项目是从9个参赛团队中挑选出来的，位于Anhanguera高速公路的边界，距离铁特河一公里。该项目紧邻公司的配送中心，占地面积约11.2万平方米，植被茂密。该项目占地29700m²，包括可容纳1600名员工的办公空间，以及辅助区域、建筑设备和公共设施。

考虑到场地的自然条件，建筑被构思为一个透明的、通透的水平大楼，大约100m长。就像一个在茂盛的植被中"漂浮"的体量，它通过穿过树梢的人行道接待往来的行人。

花园、绿地和波光粼粼的水池穿插在建筑体量中，平衡了项目的体量组成。它的内部空间由六层（一层、三个标准层和两个较低楼层）组成，围绕着一个包含所有楼层的Integrating Void（整合挑空）区。内部花园和交通流线区域也面对着这个挑空空间。全景电梯和一系列楼梯穿过空间，强化了这座建筑的外向性质，让人能在外部看到建筑使用者的运动轨迹。由水平玻璃框架和金属穿孔百叶构成的宽阔的Unifying Cover（统一覆层）可过滤自然光线。

遵循与自然共存的原则，东立面和西立面都由金属框架层压玻璃百叶窗系统提供保护。安装在南北立面的金属百叶窗进一步将室内与

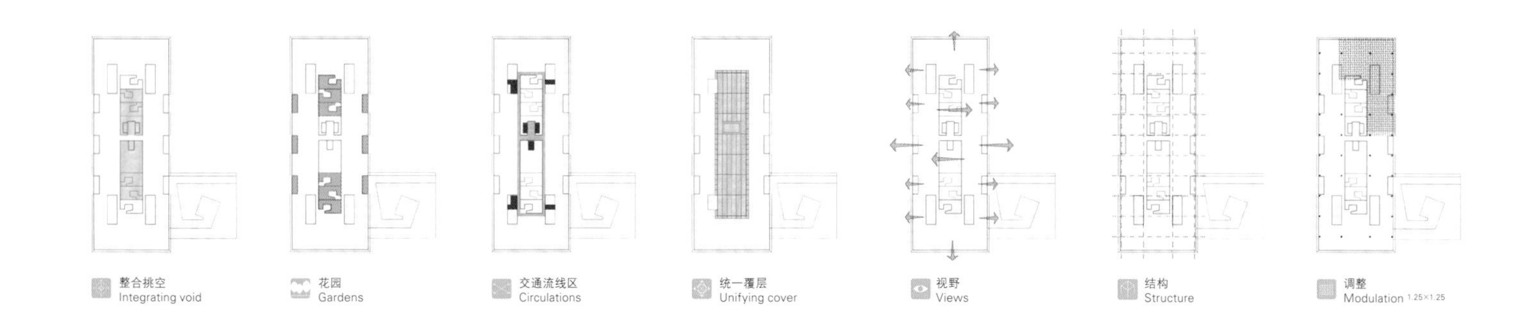

室外环境融为一体。此外，绿色屋顶旨在提高建筑的保温隔热性能。

NASP的建筑设计提供了打破传统、动态、流畅、外向的工作空间，试图将支撑、管理和驱动公司行为的原则变得更加具体化，这些原则包括可持续性、创新、透明度以及对社会环境的承诺。

Founded in 1969, Natura is a Brazilian multinational company of cosmetics and beauty products that sells products through representatives in more than 3,200 stores in 70 countries across the world.

In 2011, Natura promoted an architectural competition to design its new administrative headquarters in São Paulo. This project, chosen from among nine participating teams, is located on the borders of Via Anhanguera, one kilometer from Marginal Tietê. Built next to the company's distribution center, it occupies a plot with dense vegetation of approximately 112,000m². With an area of 29,700m², this project includes corporate spaces for 1,600 employees, as well as support areas, services and utilities.

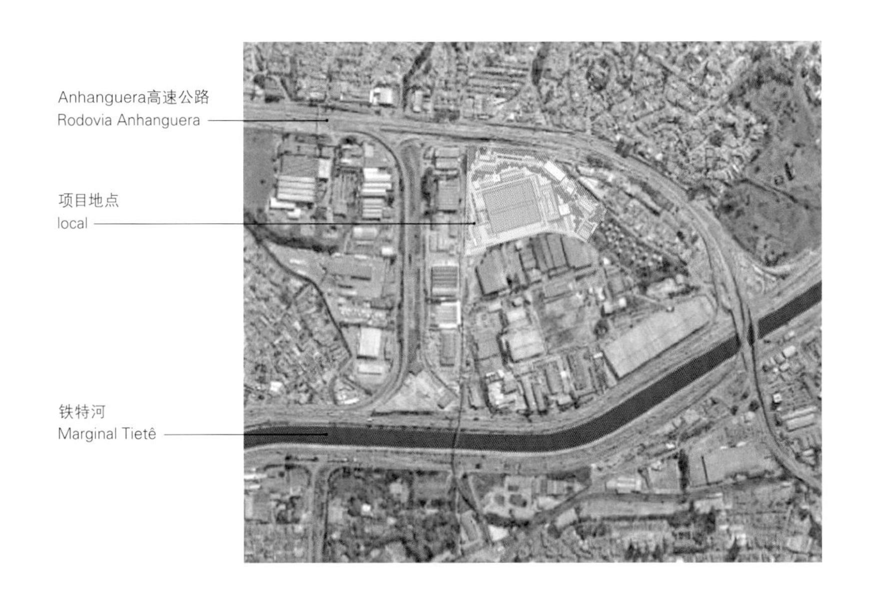

1. 主入口 2. 装卸货入口 3. 直升机停机坪 4. 辅助楼 5. 公共设施 6. 物流配送中心 7. NASP总部
1. main entrance 2. load entrance 3. helipad 4. support buildings 5. utilities 6. distribution center 7. NASP headquarters

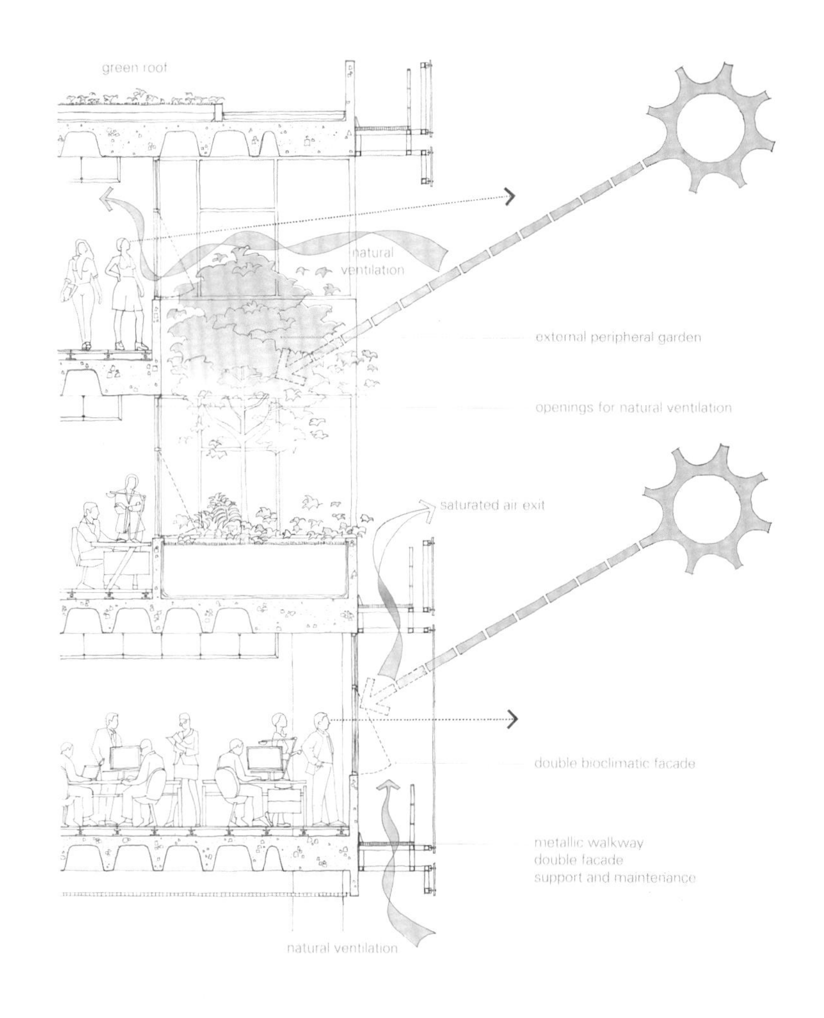

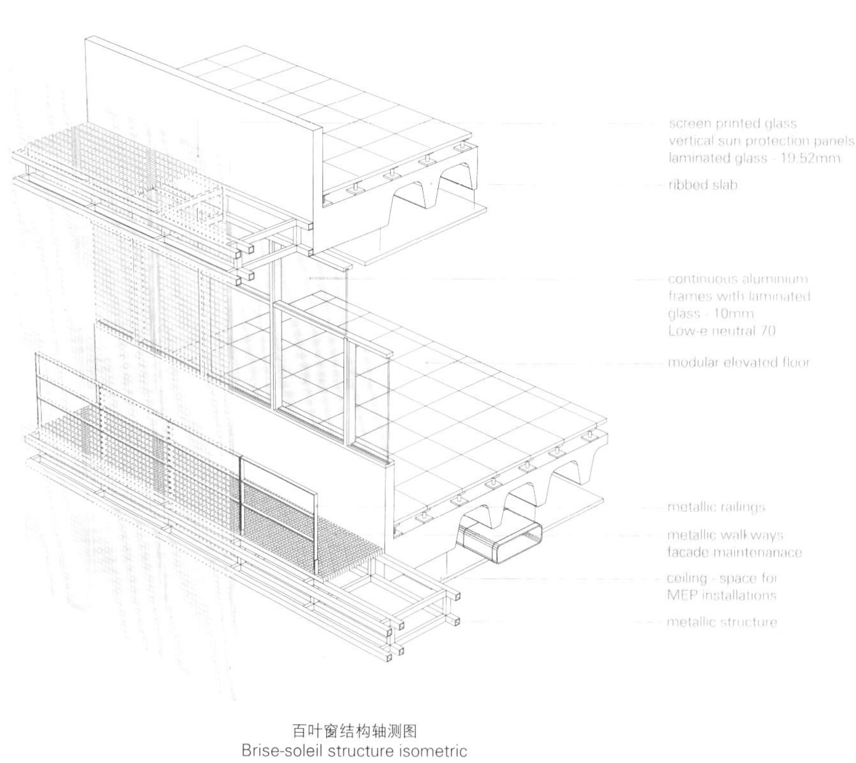

百叶窗结构轴测图
Brise-soleil structure isometric

Respecting the natural conditions of the site, the building was thought of as a transparent and highly permeable horizontal tower, approximately 100 meters long. Like a "floating" volume amid the exuberant vegetation, it receives pedestrians through walkways that pass through the treetops. Gardens, green areas and reflecting pools interject into the constructed mass and balance out its volumic composition.

Comprising six floors (ground floor, three standard stories and two lower floors), its internal spaces are articulated around an Integrating Void, which encompasses all floors. Internal gardens and circulation areas also face the void. Panoramic lifts and a set of stairs cross the space and reinforce the prerogative of an extroverted building that exposes the flow and movement of its users. A wide Unifying Cover, consisting of horizontal glazed frames and metallic perforated louvers, filters the natural light.

Following the principle of coexisting with nature, the glazed facades on the east and west are protected by a metal-framed laminated glass louver system. Metal louvers implanted on the north and south faces further integrate the interior with the outside world. Furthermore, the green roof aims to boost the building's thermal insulation.

Offering unconventional, dynamic, fluid and extroverted work spaces, NASP architecture seeks to externalize the principles that underlie, govern and drive the company's actions – sustainability, innovation, transparency and social-environmental commitment.

1. 走道
2. 全景电梯
3. 坡道
4. 办公室
5. 日光浴场
6. 花园
7. 整合挑空

1. walkway
2. panoramic lifts
3. ramp
4. offices
5. solarium
6. garden
7. integrating void

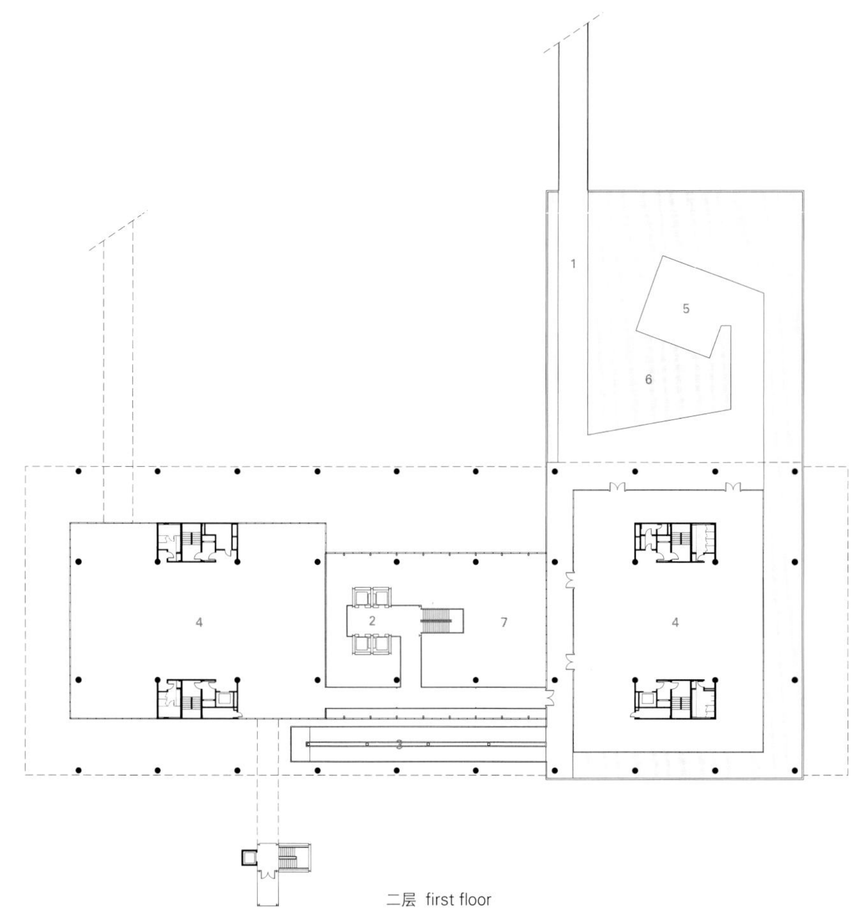

二层 first floor

1. 门廊
2. 全景电梯
3. 倒影水池
4. 餐厅
5. 厨房
6. 剧场
7. 技术区
8. 坡道
9. 停车场
10. 自行车

1. porch
2. panoramic lifts
3. reflecting pool
4. restaurant
5. kitchen
6. theater
7. technical area
8. ramp
9. parking
10. bikes

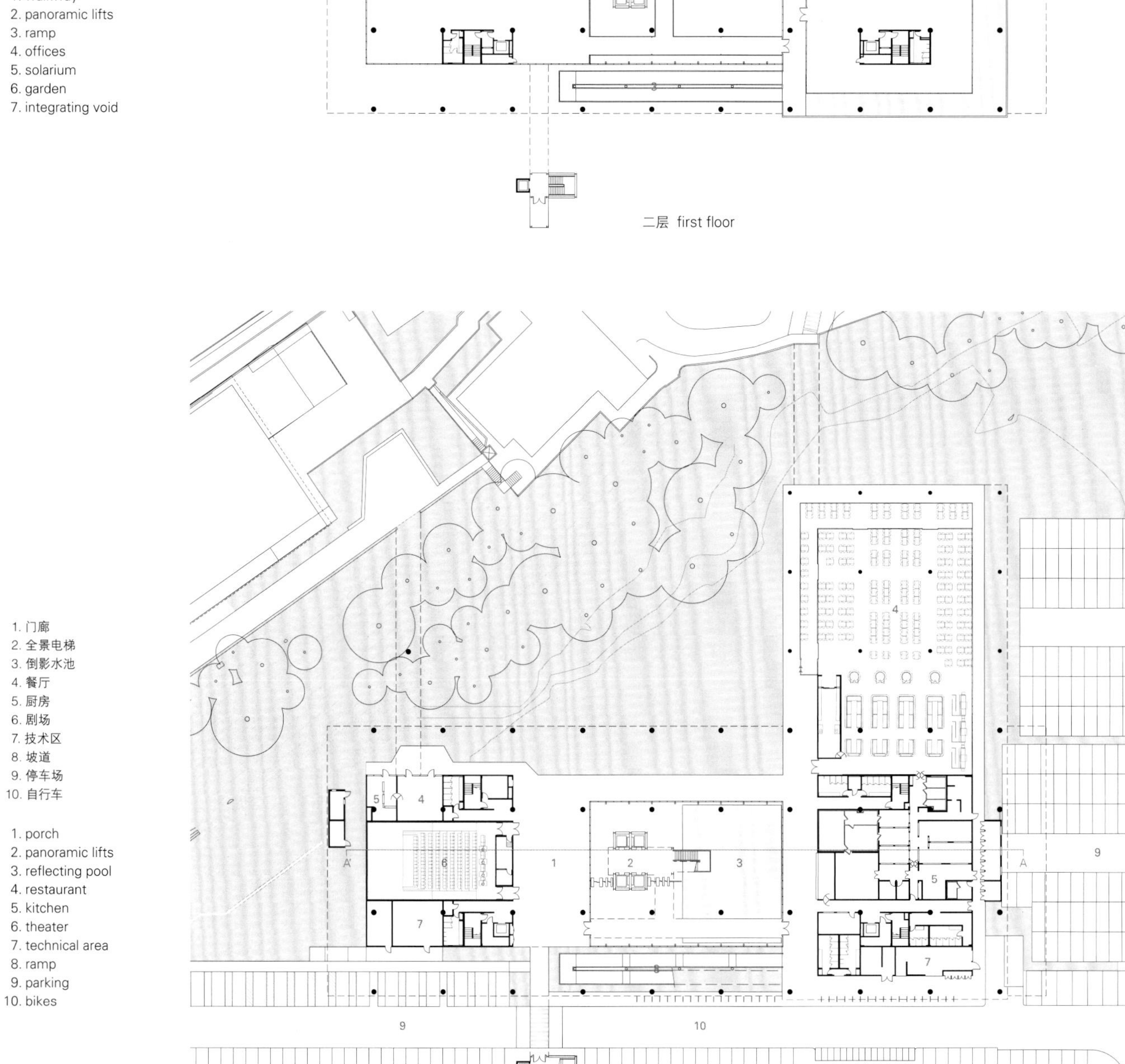

一层 ground floor

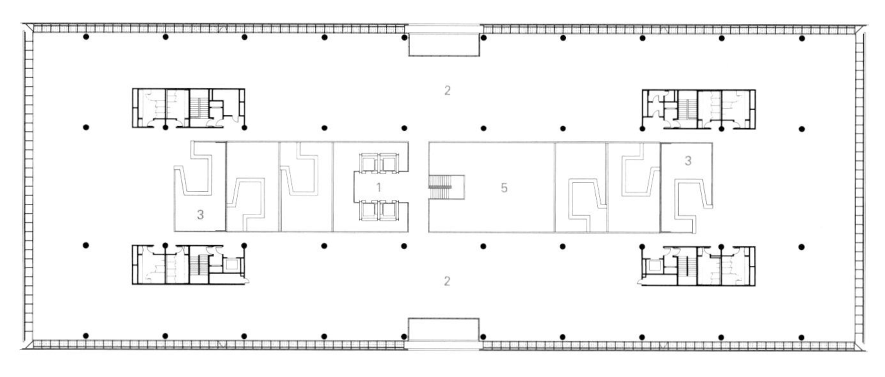

六层 fifth floor

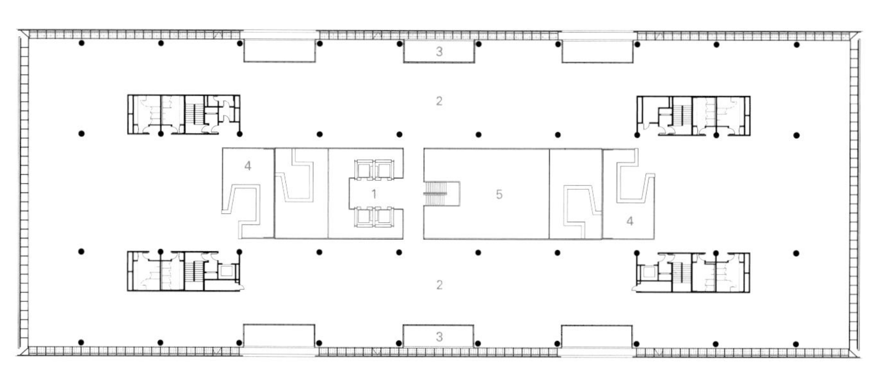

五层 fourth floor

1. 全景电梯
2. 办公室
3. 周围花园
4. 阶梯花园
5. 整合挑空

1. panoramic lifts
2. offices
3. peripheral gardens
4. stepped gardens
5. integrating void

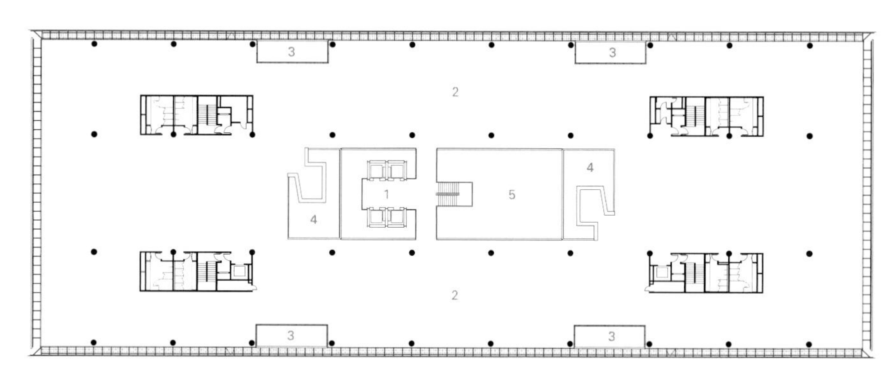

四层 third floor

1. 主入口通道
2. 配送中心通道
3. 全景电梯
4. 坡道
5. 倒影水池
6. 接待处
7. Natura商店
8. 露台
9. 整合挑空

1. main entrance access
2. distribution center access
3. panoramic lifts
4. ramp
5. reflecting pool
6. reception
7. Natura shop
8. terrace
9. integrating void

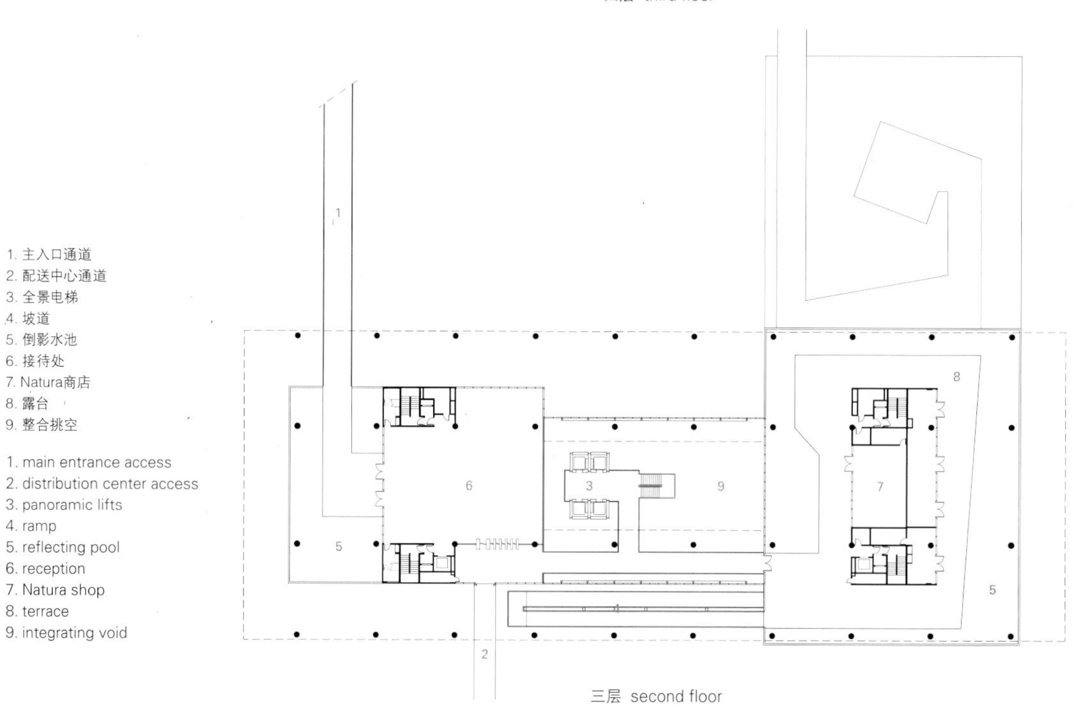

三层 second floor

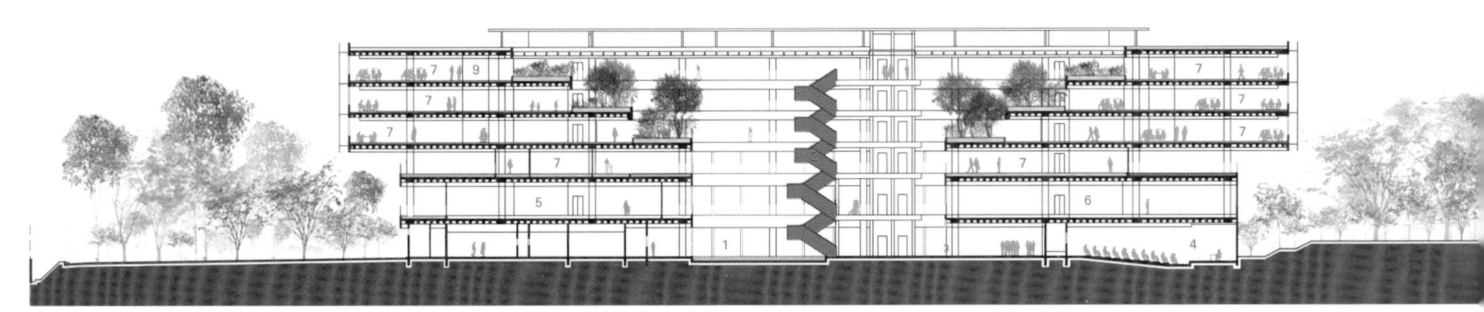

A–A' 剖面图 section A-A'

项目名称：Natura Headquarters São Paulo / 地点：São Paulo, Brasil / 建筑师：Dal Pian Arquitetos / 设计团队：Lilian Dal Pian and Renato Dal Pian - authors / 合作方：Carolina Freire - coordinator; Amanda Higuti, Bruno Pimenta, Carolina Fukumoto, Carolina Tobias, Cristina Sin, Cristiane Sbruzzi, Filomena Piscoletta, Giovana Giosa, Júlio Costa, Lídia Martello, Luís Taboada, Marina Risse, Natalie Tchilian, Paola Meneghetti, Paulo Noguer, Ricardo Rossin, Sabrina Aron and Yuri Chamon / 结构：Modus Engenharia / 基础：Damasco Penna / 设施：Tesis Engenharia / 空调设施：TR Thérmica
景观设计师：Sérgio Santana Paisagismo / 照明设计师：Franco Associados / 环境舒适：Chapman-BDSP

声效：Harmonia Acústica / 框架设计顾问：Cineplast / 建筑自动化：Crescêncio Consultoria / 可达性：Jugend Controle Predial / 防水：Pimenta Associados / 土方工程和排水：Proasp Assessoria e Projetos / 直升机停机坪：Fat's Engenharia / Helipad: Sigger / LEED认证：CTE Engenharia e Planejamento
预算：Ação Planejamento / 法律顾问：H2, Tekton, IGP, Negrisolo&Negrisolo / 管理：ARC Controle de Investimentos / 承包商：HOCHTIEF DO BRASIL
客户：Natura Cosmetics / 用地面积：111,736m² / 建筑面积：29,700m² / 设计时间：2011—2014 / 施工时间：2015—2017 / 摄影师：©Nelson Kon (courtesy of the architect) - p.174~175, p.178, p.179, p.180~181, p.184~185, p.186, p.187; ©Pedro Mascaro (courtesy of the architect) - p.170~171, p.172~173, p.176~177

默克公司创新中心

Merck Innovation Center

HENN

默克公司在达姆施塔特的场地将得到分阶段开发，将其从生产工厂改造成科技园区。这一次改造的核心是建一座创新中心，打造一个全新的工作环境。

动态空间连续体的形式产生了独立的工作场所，同时将它们连接起来，形成了一个空间网络。

该建筑面对着法兰克福特大街向后撤出了一块空间，从而产生了一个公共广场——伊曼纽尔·默克广场。建筑体量的矩形外形得自于附近建筑的大环境，同时与建筑内部运作的动态特征形成了对比。室内的特色在于不断流动的空间结构的伸展。

桥式连接体对角跨越椭圆形核心之间的空间，将各个工作区彼此连接起来。台阶、坡道和楼面呈螺旋式上升。从一个工作组到另一个工作组，从一个楼层到另一个楼层，它们之间的路线几乎是在不知不觉中毫不费力地完成的。

交叉的连桥使建筑的中心点更加密集，并将工作站上方6m的空间高度降低到3m，看起来像是漂浮在水面上。建筑荷载由立面支撑和仅仅四根室内柱子所承担。由于外部覆层为高度抛光的不锈钢板，所以这些柱子看上去仿佛消失于无形。

每一层都有两个工作区，彼此对角布置。每个工作空间都是为由外部和内部工作人员组成的项目组而设计的，大家会在临时项目或原有项目的基础上进行合作创新。集中工作室和会议室沿着立面和夹层布置。一层包括咖啡厅、休息室和礼堂；二层则为图书馆和开放式工作站；顶层可容纳一个工作坊。

采用跨度可达20m的钢筋混凝土组合结构，可使工作区域实现无柱布置。

立面设置在开放的外部板条后面，这些板条以不同的角度铺设，反过来在外部呈现了一个动态的外观，并使室内空间的双层特征从外部看上去更加清晰。在日光和声学方面，全方位的层高透明外墙和吸声效果极佳的天花板提供了理想的工作条件。此外，创新中心在照明、装修和天窗方面采用了默克公司的一系列新产品和新技术。最新的OLED（有机发光二极管）技术应用于"光云"艺术装置，以及"媒体墙"监控装置。

创新中心通过开放式楼梯与员工餐厅相连，餐厅采用了创新中心弯曲流动的建筑设计语汇。员工餐厅由一层的美食广场和二层的餐厅设施组成。螺旋楼梯、椭圆形柜台、建筑中心的浓缩空间和角落的宽敞开放空间都与创新中心的设计相互呼应。

这两座建筑都通过了LEED白金标准认证。

Merck's site in Darmstadt is to be developed stage-by-stage, remodelling it from a production works into a technology and science campus. The heart of this transformation is the Innovation Center with a new world of work.
A dynamic spatial continuum singularises the individual

项目名称：Merck Innovation Center / 地点：Emanuel-Merck-Platz 1, Darmstadt, Germany / 建筑师：HENN / 项目管理：Drees & Sommer, Frankfurt am Main
结构工程师：Bollinger + Grohmann Ingenieure (Berlin) / 供暖、通风、卫浴、制冷、测量与控制技术：ZWP Ingenieur - AG (Wiesbaden)
立面：Emmer Pfenninger Partner AG (Münchenstein, Schweiz) / 厨房：IGW Ingenieurgruppe Walter (Stuttgart) / 室外设施：Topotek 1 (Berlin)
媒体技术：macom GmbH, Offenbach a. M. / 消防：Hamann Ingenieure GmbH (Berlin) / 建筑物理和声学：Müller BBM (Berlin)
照明设计：Lumen 3 GbR (München) / 照明设计II：W-tec AG (Bad Homburg)

workplaces whilst connecting them to form a spatial network.

The building is set back facing Frankfurter Straße, thus generating the space for a public square – Emanuel Merck Platz. The orthogonal shape of the architectural volume is derived from the context of the neighbouring buildings, simultaneously acting as a contrast to the animation of the building's inner workings. The interior is characterised by the unfolding of a continuously flowing spatial structure.

Bridge-like connections diagonally span the space between the oval cores, linking the individual workspaces with each other. Steps, ramps and floor areas spiral upwards. The routes between one workgroup and another, from one level to the next, are accomplished almost imperceptibly and effortlessly.

The inter-crossing bridges, which densify the center point of the building and diminish the spatial height of 6 meters to 3 meters above the workstations, appear to float. The strain of the loads is absorbed by supports along the facade and a mere four interior columns. Due to their highly polished

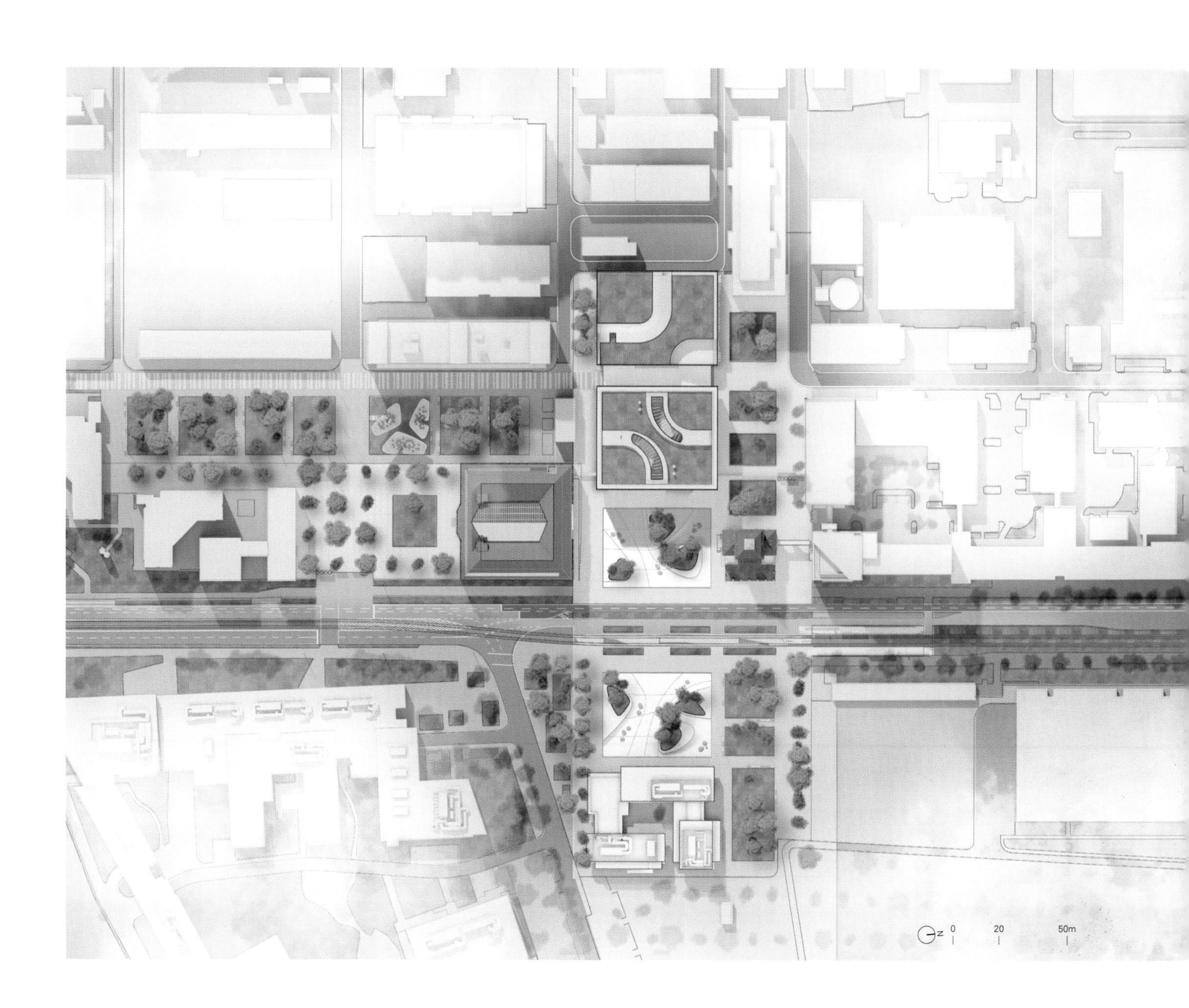

标识与标签：moniteurs GmbH, Kommunikationsdesign / “媒体墙”装置：Art + Com AG (Berlin) / “光云”装置：Tamschick Media + Space (Berlin)
客户：Merck KGaA / 用途：office/research building and staff restaurant / 总建筑面积：Innovation center - 12,500m²; Staff restaurant - 9,500m²
高度：Innovation center - 27.70m; Staff restaurant - 19.10m / 工作空间创新中心：450 / 员工餐厅：1,000 seats / 认证：LEED Platin
建筑规模：Innovation center - one story below ground, five stories above ground; Staff restaurant - one story below ground, two stories above ground
施工时间：2015.10—2018.2 / 摄影师：©HGEsch Photography (courtesy of the architect)

stainless steel coverings, the columns have a practically dematerialized presence.

Every level has two work areas, positioned diagonally vis-à-vis one another. Each workspace is designed to provide for a project group consisting of external and internal staff, cooperating together on innovations either on a temporary or a project basis. Concentration and meeting rooms are arranged along the facade and on the mezzanines. The ground floor contains a café, a lounge and an auditorium; the first upper storey a library and open workstations; the top floor accommodates a workshop.

By using a reinforced concrete composite construction with spans of up to 20 meters, the work areas can be laid out column free.

The facades are set behind open external slats that run at varying angles, in turn giving the exterior a dynamic appearance and making the double-storey character of the interior spaces legible from the outside. The all-round storey-high transparent facades and the highly sound-absorbent ceilings provide ideal working conditions in terms of daylight and acoustics. In addition, the Innovation Center features a whole series of new Merck products and technologies in the lighting, the finishings and the skylight. The latest OLED technology was applied in the Light Cloud art installation, as well as the Media Wall monitor installation.

The Innovation Center is connected to a staff restaurant via an open stairway, the restaurant adopting the curved and flowing architectural vocabulary of the Innovation Center. The staff canteen consists of a food court on the ground floor with the restaurant facilities on the two upper floors. Spiral staircases, oval counters, the condensed space in the center of the building and the amplified, open space at the corners all echo the Innovation Center.

Both buildings are certified with the LEED's platinum standard.

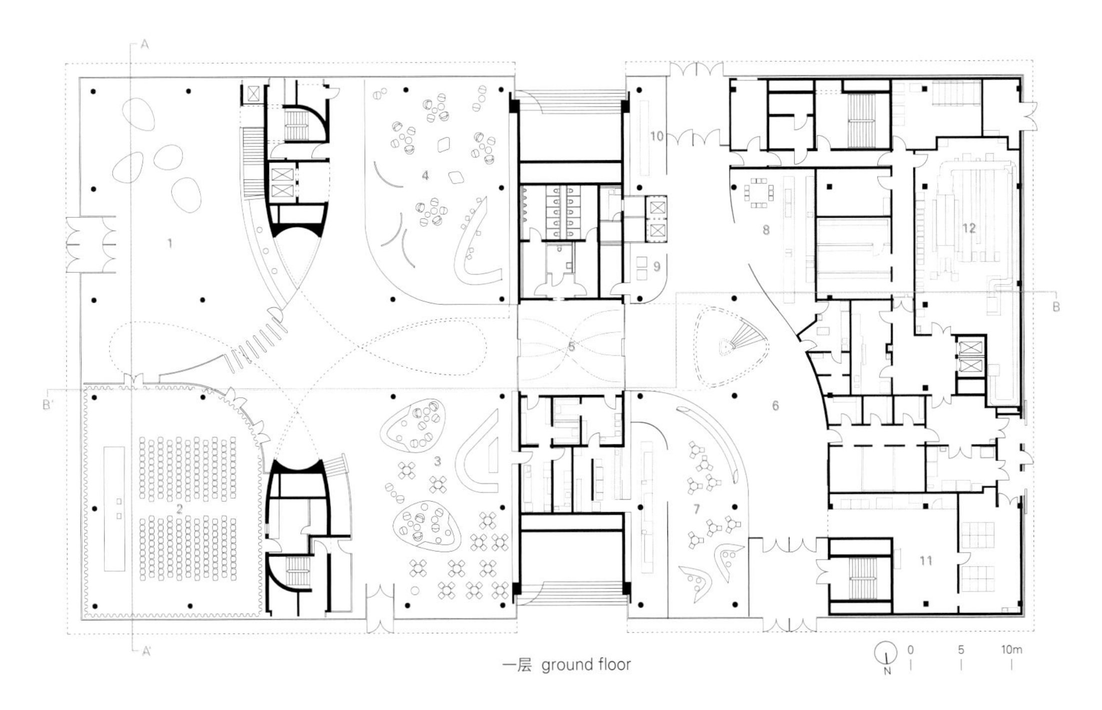

一层 ground floor

1. 门厅
2. 礼堂
3. 咖啡厅
4. 休息室
5. “光云”装置
6. 美食花园
7. 阳台
8. 酒吧与餐厅
9. IT商店
10. 推广空间
11. 饮料店
12. 厨房

1. foyer
2. auditorium
3. cafe
4. lounge
5. light cloud
6. food court
7. balcony
8. bar & restaurant
9. IT shop
10. promotion space
11. beverage shop
12. kitchen

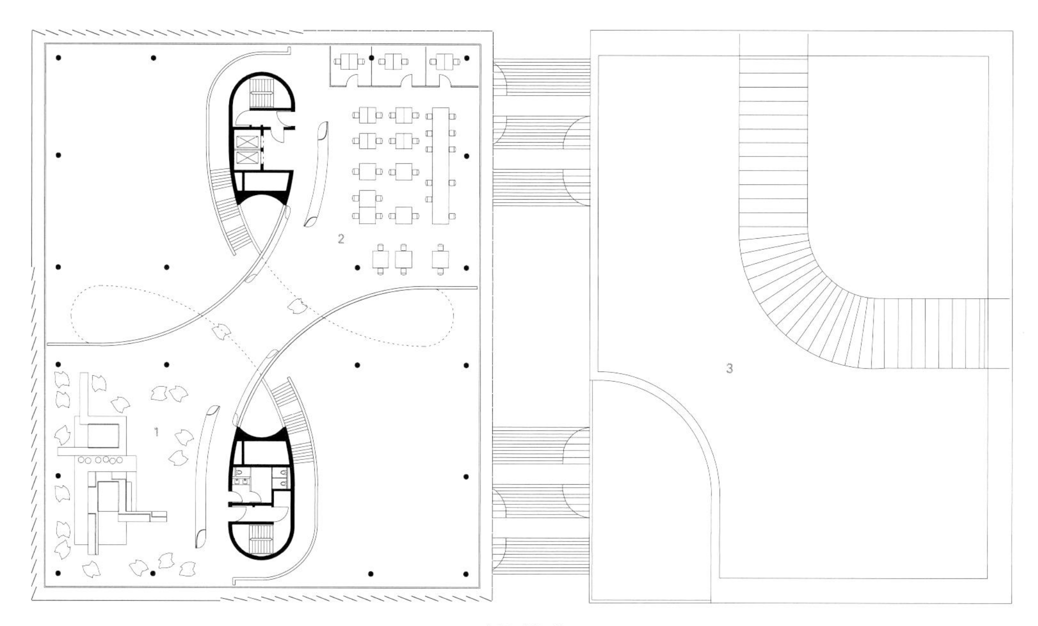

1. 重新创作区
2. 创客空间
3. 屋顶

1. re-creation
2. maker space
3. roof top

六层 fifth floor

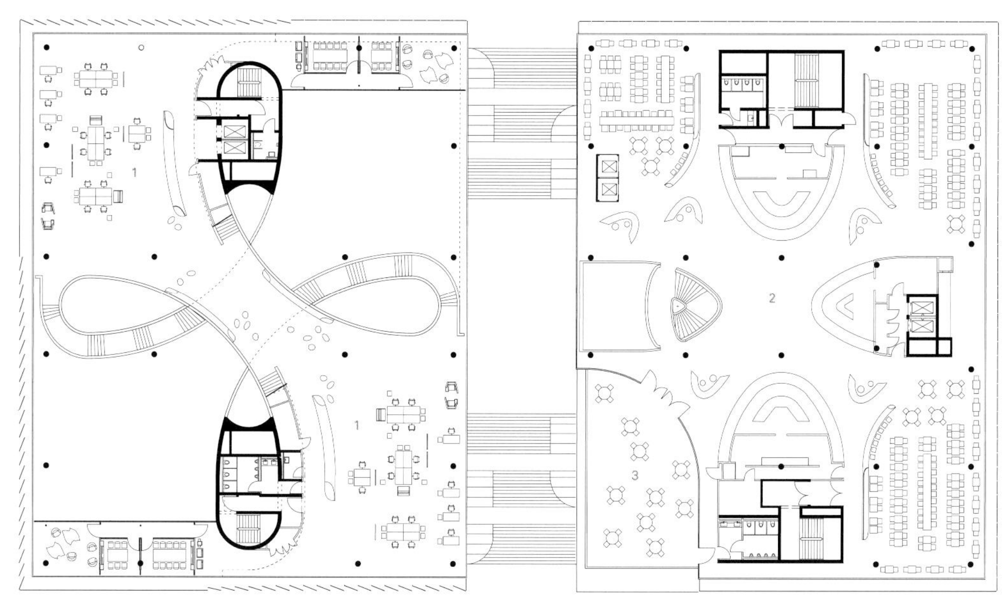

1. 工作空间
2. 员工餐厅
3. 露台

1. working space
2. staff canteen
3. terrace

三层 second floor

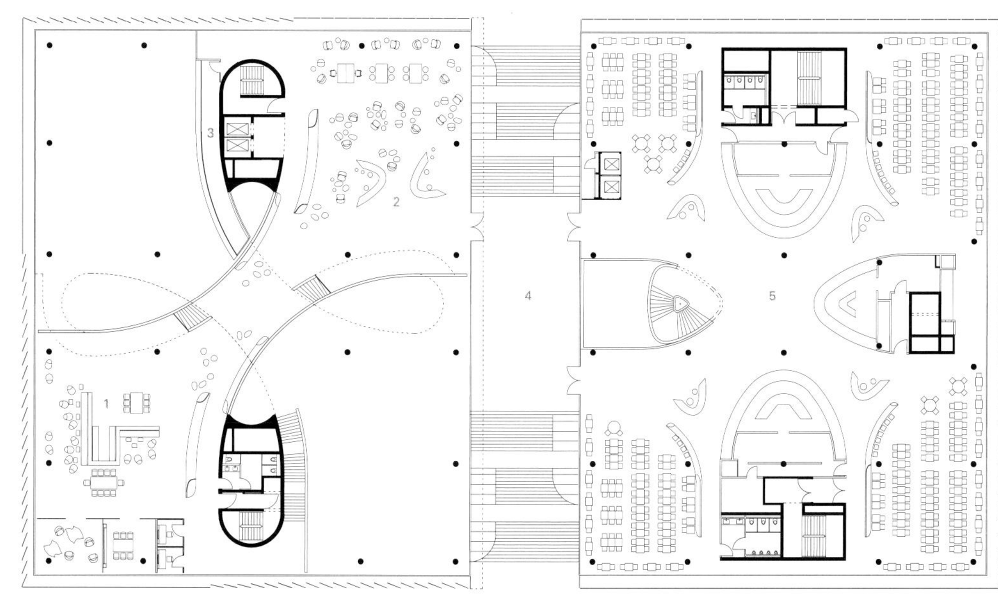

1. 共同创作区
2. 图书馆
3. “媒体墙”装置
4. 楼梯
5. 员工餐厅

1. co-creation
2. library
3. media wall
4. staircase
5. staff canteen

二层 first floor

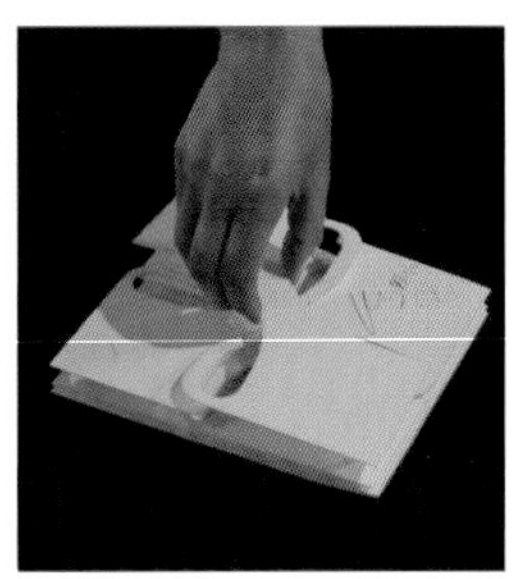

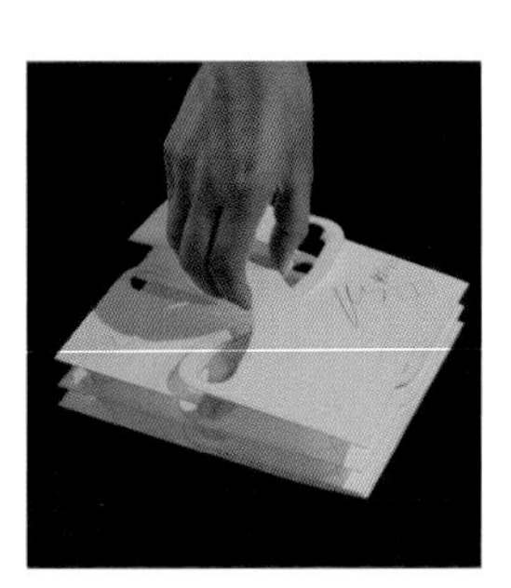

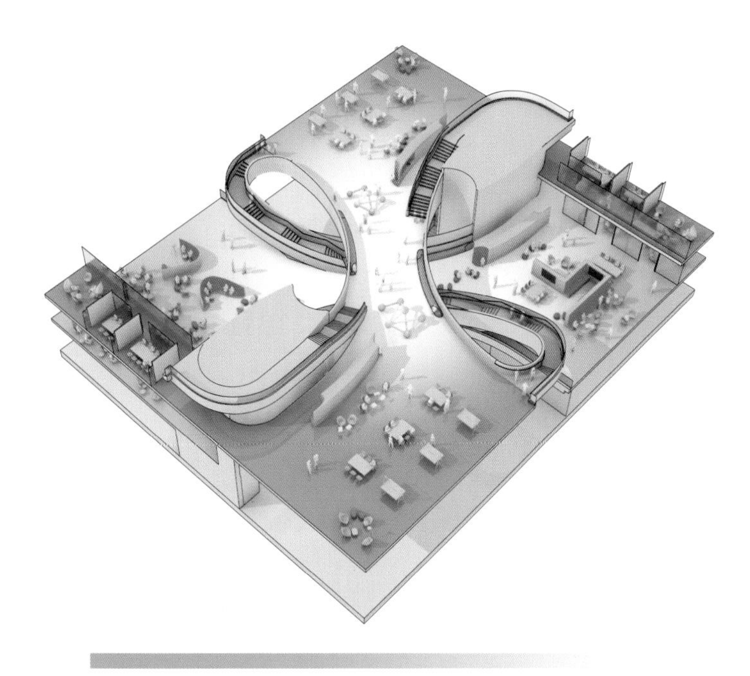
专心工作区
concentration
交流区
communication

MERCK

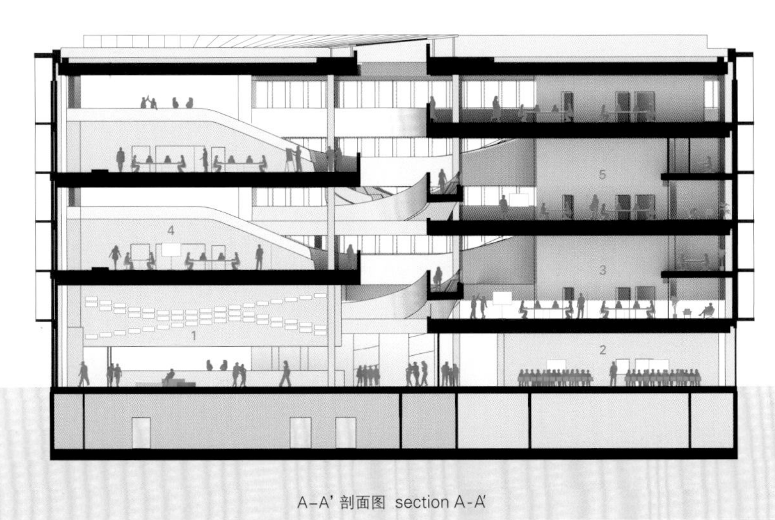

1. 门厅	1. foyer
2. 礼堂	2. auditorium
3. 共同创作区	3. co-creation
4. 工作空间	4. working space
5. 重新创作区	5. re-creation
6. 厨房	6. kitchen
7. 酒吧与餐厅	7. bar & restaurant
8. “光云” 装置	8. light cloud
9. 咖啡厅	9. cafe
10. 员工餐厅	10. staff canteen
11. 楼梯	11. staircase
12. 屋顶	12. roof top

A-A' 剖面图 section A-A'

B-B' 剖面图 section B-B'

©Ingrid von Kruse

P110 David Chipperfield Architects

David Chipperfield was born in London, England in 1953 and graduated from Kingston School of Arts in 1976. He received his architecture diploma at Architectural Association (AA) in 1977. After graduation, he worked for several notable architects such as Norman Foster and Richard Rogers, and established David Chipperfield Architects in 1985. The firm builds its reputation from meticulous attention to the concept and details of every project, and a relentless focus on refining the design ideas to arrive at a solution which is architecturally and socially coherent. Develops a diverse international works including cultural, residential, commercial, leisure and civic projects. Won more than 100 international awards and citations for design excellence, including the RIBA Stirling Prize in 2007, and the European Union Prize for Contemporary Architecture – Mies van der Rohe Award, and the Deutscher Architekturpreis in 2011.

P130 BCHO Architects

Was opened by Cho Byoung-soo in 1994. He has been actively practicipating with themes such as 'experience and perception', 'existing and existed', 'I shaped house and 'L shaped house', 'contemporary vernacular', 'organic versus abstract'. Has taught the theory and design of architecture at several universities including Harvard University, Columbia University, University of Hawaii and Aarhus University. Is the recipient of Architizer A+ Awards and AIA Northwest and Pacific Region Design Awards in 2015, KIA Award in 2014 with several previous AIA Honor Awards in Montana Chapter and in N.W.Pacific Regional. His projects has been published in different magazines including The Architectural Reveiw, Dwell, and deutsche bauzeitung(db).

P52 Herzog & de Meuron

Was established in Basel, 1978. Has been operated by senior partners; Christine Binswanger, Ascan Mergenthaler and Stefan Marbach, with founding partners Pierre de Meuron and Jacques Herzog. An international team of about 40 Associates and 380 collaborators is working on projects across Europe, the Americas and Asia. The firm's main office is in Basel with additional offices in Hamburg, London, New York City, and Hong Kong. The practice has been awarded numerous prizes including the Pritzker Architecture Prize (USA) in 2001, the RIBA Royal Gold Medal (UK) and the Praemium Imperiale (Japan), both in 2007. In 2014, awarded the Mies Crown Hall Americas Prize (MCHAP).

P188 HENN

Gunter Henn was born in 1947 in Dresden, Germany. Studied Architecture and Civil Engineering at TU Berlin, ETH Zurich and TU Munich, where he also was awarded a PhD in Architecture. Established his architectural studio HENN in Munich, 1979. HENN now has offices in Munich, Berlin and Beijing. Managing Director and Head of Design, Martin Henn undertook his architectural studies at the University of Stuttgart and ETH Zurich, where he obtained his master's degree in 2006. Received a Master of Advanced Architectural Design at Columbia University in New York in 2008. Prior to beginning at HENN, he was employed at Zaha Hadid Architects in London and Asymptote Architecture in New York. Has been working for the HENN office since 2008.

Silvio Carta

Is an architect and researcher based in London. Received Ph.D. from University of Cagliari, Italy in 2010. His main fields of interest is architectural design and design theory. His studies have focused on the understanding of the contemporary architecture and the analysis of the design process.Taught at the University of Cagliari, Willem de Kooning Academy of Rotterdam and Delft University of Technology. He is now senior lecturer at the University of Hertfordshire. Since 2008 he has been editor-at-large for C3, and his articles have also appeared in A10, Mark, Frame and so on.

P72 Foster + Partners

Is an international studio for architecture, engineering and design, led by Founder and Chairman Norman Foster and a Partnership Board. Founded in 1967, the practice is characterized by its integrated approach to design, bringing together the depth of resources required to take on some of the most complex projects in the world. Over the past five decades the practice has pioneered a sustainable approach to architecture and ecology through a strikingly wide range of work, from urban masterplans, public infrastructure, airports, civic and cultural buildings, offices and workplaces to private houses and product design. The studio has established an international reputation with buildings such as the world's largest airport terminal at Beijing, Swiss Re's London Headquarters, Hearst Headquarters in New York, Millau Viaduct in France, the German Parliament in the Reichstag, Berlin, The Great Court at London's British Museum, Headquarters' for HSBC in Hong Kong and London, and Commerzbank Headquarters in Frankfurt.

Heidi Saarinen

Is a London based designer, lecturer at Coventry University and also an artist with current research focused on space and place. Is interested in the peripheral space, in-between and the interaction and collision between architecture, spaces, city, performance and the body. Is currently working on a series of interdisciplinary projects linking architecture, heritage, film and choreography in the urban environment. Is part of several community and creative groups in London and the UK where she engages in events and projects highlighting awareness of community and architectural conservation in the built environment.

P154 BIG

Was founded in 2005 by Bjarke Ingels and now based in Copenhagen and New York. The group of architects, designers, builders, and thinkers operates within the fields of architecture, urbanism, interior design, landscape design, product design, research and development. The office is currently involved in a large number of projects throughout Europe, North America, Asia and the Middle East. Their architecture emerges out of a careful analysis of how contemporary life constantly evolves and changes.

P170 Dal Pian Arquitetos

Was founded in São Paulo, Brazil in 1992 by Lilian Dal Pian[right] and Renato Dal Pian[left]. Is a member of AsBEA – Brazilian Association of Architecture Offices. Lilian and Renato have started their professional collaboration in 1981 and developed numerous activities in London and Milan from 1987 to 1992. Lilian Dal Pian graduated in 1981 and received a Master's degree in 2004 from the Department of Architecture and Urbanism, University of São Paulo[FAUUSP]. Renato Dal Pian graduated from the Department of Architecture and Urbanism, Pontifical Catholic University of Campinas[FAUPUCC] in 1981 and received a Master's degree from the Department of Architecture and Urbanism, Mackenzie Presbyterian University[FAUMACK] in 2002. Has been a Professor of Project Department at the FAUMACK since 1993.

©Michael Powers

P90 Morphosis

Founded in 1972, the firm is committed to the practice of architecture as a collaborative enterprise, with founder and Pritzker-prize winning architect Thom Mayne serving as design director alongside principals Arne Emerson, Ung-Joo Scott Lee, Brandon Welling, and Eui-Sung Yi, and more than 60 professionals working in Los Angeles, New York, and Shenzhen. Born in Connecticut, Thom Mayne studied architecture at the University of Southern California and Harvard GSD. Was a founding member of the SCI-Arc in 1972. Was honored with the Pritzker Prize in 2005, AIA Los Angeles Gold Medal in 2000 and the AIA Gold Medal in 2013. Was appointed to the President's Committee on the Arts and Humanities in 2009. Has held teaching positions at Columbia, Yale (the Eliel Saarinen Chair in 1991), the Harvard GSD (Eliot Noyes Chair in 1998), the Berlage Institute, the Bartlett School of Architecture, and many other institutions around the world. Is a tenured Professor at UCLA Architecture and Urban Design since 1993.

P18 3r Ernesto Pereira

Ernesto Pereira is a Portuguese architect in Vila Chã, Porto. Received a Master's degree from the Lusiada University of Porto in 2009 and began his professional career in the studio of the architect Alvaro Leite Siza Vieira. Opened his own atelier 3r Ernesto Pereira in 2011. Is passionate about architecture and idealizes it. Believes the artistic inspiration comes from the most varied and basic things of life. Prefers working with strong concepts that convey the identity and the character of not only the architect but also the client.

P10 Llama Urban Design

Is an architecture and urban design studio, founded by Mariana Leguía[left] and Angus Laurie[right]. Mariana studied architecture at the University of Ricardo Palma in Lima. Angus studied economics at the University of King's College, Canada. Both hold a master's degree in City Design from the London School of Economics. After working in diverse design studios in the USA and the United Kingdom, they founded LLAMA in Lima, Peru in 2010. LLAMA was nominated for the Mies Crown Hall Americas Prize for Emerging Architecture in 2018. Was selected to represent Peru in the 2017 Latin-American Architecture Biennale[BAL] in Pamplona, Spain. In 2016, won the residential category of the Ontario Wood Design Awards.

P10

P36 Felipe Rodrigues

Is a Brazilian architect with more than 17 years of experience as a project coordinator and an architect. Received his degree from The School of Architecture and Urbanism of Mackenzie Presbyterian University in 1997. After graduation, he has worked at Calder Architecture & Management and Figueiredo Ferraz Consulting & Project Engineering. Has been managing his own office, Felipe Rodrigues Arquitetos as Director and Partner since 2003.

P28 Life Style Koubou

Is a Japanese architectural practice led by Kotaro Anzai. He has been always impressed by the four beautiful seasons of Mt Adatara. Having grown up here, surrounded by Mt Adatara, he learned the wisdom of living. This valuable wisdom, which has been passed through the generations, has both eternal and universal value. To realize and develop these wisdom and values, he carefully examines both the climate and nature of a specific area. After that, he proposes ideas that locals love and support.

图书在版编目(CIP)数据

标识与身份 : 汉英对照 / 瑞士赫尔佐格和德梅隆建筑事务所等编 ; 贾子光, 周荃译. -- 大连 : 大连理工大学出版社, 2019.7

(建筑立场系列丛书)

ISBN 978-7-5685-2076-8

Ⅰ. ①标… Ⅱ. ①瑞… ②贾… ③周… Ⅲ. ①建筑设计—汉、英 Ⅳ. ①TU2

中国版本图书馆CIP数据核字(2019)第124513号

出版发行：大连理工大学出版社
（地址：大连市软件园路80号　　邮编：116023）
印　　刷：上海锦良印刷厂有限公司
幅面尺寸：225mm×300mm
印　　张：13
出版时间：2019年7月第1版
印刷时间：2019年7月第1次印刷
出 版 人：金英伟
统　　筹：房　磊
责任编辑：张昕焱
封面设计：王志峰
责任校对：杨　丹
书　　号：978-7-5685-2076-8
定　　价：258.00元

发　行：0411-84708842
传　真：0411-84701466
E-mail：12282980@qq.com
URL：http://dutp.dlut.edu.cn